John Mason

Self-Knowledge

John Mason

Self-Knowledge

ISBN/EAN: 9783337258696

Printed in Europe, USA, Canada, Australia, Japan

Cover: Foto ©berggeist007 / pixelio.de

More available books at **www.hansebooks.com**

A TREATISE,

SHEWING THE

NATURE and BENEFIT

OF THAT

IMPORTANT SCIENCE,

AND

The WAY to attain it.

INTERMIXED

With various Reflections and Observations on HUMAN NATURE.

By JOHN MASON, *A. M.*

―――― e Cœlo descendit γνωθι σεαυτον. Juv.
The proper Knowledge of Mankind is MAN. Pope.

THE EIGHTH EDITION.

LONDON:
Printed for James Buckland, at the Buck in Paternoster Row; and E. and C. Dilly, in the Poultry.
MD.CC.LXIX.

TO

SAMUEL LESINGHAM, *Esq*;

Treasurer of *St. Thomas's Hospital.*

SIR,

MODESTY and Self-diffidence are the allowed Characteristicks of *Self-Knowledge.* If then my presuming to address this Piece to you may seem to discover more Assurance and Self-confidence than becomes a true Acquaintance with the Subject I write upon, I have only this to say; your known Condescension and
Candour

Candour have encouraged that Presumption: Nor can any Thing animate an Address of this Nature more, than an Assurance that the Person to whom it is made, has so good an Understanding in the practical Part of this Subject, as will incline him to excuse the Defects that may appear in the Management of it.

But after all, SIR, my own Proficiency in this Science is so poor, that I dare not be confident I am not wrong in my Views, with which I desire this small Tract may appear under your Patronage. That it may have a Refuge from the Petulence of Censure, an Encouragement in the Publication, and I, at the same Time, an Opportunity of testifying my grateful Sense of many past Favours, are my open and avowed Ends herein. But still, whether an Ambition to be known to the World under the Advantage of your Friend-

DEDICATION.

Friendship be not the secret and true Motive, I cannot be certain.

However, if in this Point I may be mistaken, there is another in which I think I cannot; and that is, that it is at least a pardonable Ambition; in which I shall certainly stand acquitted by every one who knows your Character, the Delicacy of your Taste in the Choice of Friends, and the real Honour it does to those you are pleased to admit into that Number.

But even this, SIR, your Penetration will soon discover to proceed from the same Vanity I before suspected myself to be guilty of. And the World will judge, that I speak it rather to do *myself* Honour than *you*. However, I am beforehand with them in the Observation. And that I may not be tempted, in this Address, to enhance your Character (according to the usual

Stile

Stile of Dedications) in order to do Honour to my own, and at once oppress your Modesty and expose my Vanity, I shall put an End to it, without so much as attempting to describe a Character, which I shall, however, always aim to imitate.

But that you may continue to adorn that publick and useful Station you are in, and long live a Patron and Pattern of solid and disinterested Virtue; and that your many charitable Offices, and good Works on Earth, may meet with a large and late Reward in Heaven, is the hearty Prayer of,

SIR,

Your much obliged, and

very humble Servant,

Dorking, Jan. 31, 1744-5.

J. MASON.

THE
PREFACE.

THE Subject of the ensuing Treatise is of great Importance; and yet I do not remember to have seen it cultivated with that Precision, Perspicuity and Force, with which many other Moral and Theological Themes have been managed. And indeed it is but rarely that we find it professedly and fully recommended to us, in a set and regular Discourse, either from the Pulpit or the Press. This Consideration, together with a full Persuasion of its great and extensive Usefulness, hath excited the present Attempt, to render it more familiar to the Minds of Christians.

Mr. *Baxter* indeed has a Treatise upon this Subject; intitled, *The Mischief of Self-ignorance, and the Benefit of Self-acquaintance.* And I freely acknowledge some Helps I received from him. But he hath handled it (according to his Manner) in so lax and diffuse

diffuse a Way, introducing so many Things into it that are foreign from it, omitting others that properly belong to it, and skimming over some with a too superficial Notice, that I own I found myself much disappointed in what I expected from him. And was convinced that something more correct, nervous and methodical was wanting on this Subject.

I am far from having the Vanity to think that this, which I now offer to the Publick, is intirely free from those Faults which I have remarked in that pious and excellent Author; and am sensible, that if I do not fall under a much heavier Censure myself, it must be owing to the great Candour of my Reader; which he will be convinced I have some Title to, if he but duly consider the Nature and Extent of the Subject. For it is almost impossible to let the Thoughts run freely upon so copious and comprehensive a Theme, in order to do Justice to it, without taking too large a Scope in some Particulars that have a close Connexion with it: as I fear I have done (Part I. Chap. XIV.) concerning the Knowledge, Guard and Government of the *Thoughts*.

But there is a great Difference between a short, occasional and useful *Digression*, and a wide *Rambling* from the Subject, by following the Impulse of a luxuriant Fancy. A judicious Taste can hardly excuse the latter; though

though it may be content the Author should gather a few Flowers out of the common Road, provided he soon returns into it again.

This brings to my Mind another Thing, for which, I am sure, I have great Reason to crave the Reader's Indulgence; and that is, the free Use I have made of some of the *antient Heathen Writers* in my marginal Quotations, which I own looks like an Ostentation of Reading, which I always abhorred. But it was conversing with those Authors that first turned my Thoughts to this Subject. And the good Sense I met with in most of their Aphorisms and Sentiments, gave me an Esteem for them; and made it difficult for me to resist the Temptation of transcribing several of them, which I thought pertinent to the Matter in Hand. But after all, I am ashamed to see what an old-fashioned Figure they make in the Margin. However, if the Reader thinks they will too much interrupt the Course of the Subject, he may intirely omit them: tho' by that means he will perhaps lose the Benefit of some of the finest Sentiments in the Book.

I remember a modern Writer, I have very lately read, is grievously offended with Mr. *Addison* for so much as mentioning the Name of *Plato*, and presuming in one of his *Spectators* to deliver his Notions of *Hu-*

mour in a Kind of Allegory, after the Manner of that *Greek* Author; which he calls a *formal Method of trifling, introduced under a deep Ostentation of Learning, which deserves the severest Rebuke* (a). And perhaps a more severe one was never given upon so small a Provocation. From Gentlemen of so refined and delicate a Taste I can expect no Mercy. But the Publick is to judge, whether this be not as culpable an Affectation as the contrary one, which prevailed so much in the last Century.

One great View I had in mine Eye when I put these Thoughts together, was the Benefit of Youth, and especially those of them that are Students and Candidates for the Sacred Ministry; for which they will find no Science more immediately necessary (next to a good Acquaintance with the Word of GOD) than that which is recommended to them in the following Treatise; to which every Branch of human Literature is *subordinate*, and ought to be *subservient*. For certain it is, the great End of Philosophy, both Natural and Moral, is to *know ourselves*, and to *know GOD. The highest Learning is to be wise, and the greatest Wisdom is to be good*; as *Marcus Antoninus* somewhere observes.

It has often occurred to my Mind in digesting my Thoughts upon this Subject,

what

(a) See *Introduction to an Essay towards fixing the true Standard of Wit*, &c. pag. 20, 21.

what a Pity it is that this most useful Science should be so generally neglected in the modern Methods of Education; and that Preceptors and Tutors, both in publick and private Seminaries of Learning, should forget that the *forming* the *Manners* is more necessary to a finished Education than *furnishing* the *Minds* of Youth. *Socrates, who made all his Philosophy subservient to Morality* (*b*), was of this Sentiment; and took more Pains to rectify the Tempers, than replenish the Understandings of his Pupils; and looked upon all Knowledge as useless Speculation, that was not brought to this End, to make us *wiser* and *better* Men. And without doubt, if in the Academy the Youth has once happily learned the great Art of managing his Temper, governing his Passions, and guarding his Foibles, he will find a more solid Advantage from it in After-life, than he could expect from the best Acquaintance with all the Systems of ancient and modern Philosophy.

It was a very just and sensible Answer, which *Agesilaus*, the *Spartan* King, returned to one who asked him, *What it was in which Youth ought principally to be instructed?* He replied, *that which they have most need to practise when they are Men* (*c*). Were this single Rule

---

(*b*) Totam philosophiam revocavit ad mores. *Sen. Epist.* 72.

(*c*) See *Plutarch*'s *Laconic* Apothegms under the Word *Agesilaus*.

Rule but carefully attended to in the Method of Education, it might probably be conducted in a Manner much more to the Advantage of our Youth than it ordinarily is. For as Dr. *Fuller* observes, *that Pains we take in Books or Arts, which treat of Things remote from the Use of Life, is but a busy Idleness* (d). And what is there in Life which Youth will have more frequent Occasion to practise than this? What is there which they afterwards more regret the Want of? What is there in which they want more Direction and Assistance than the right Government of their Passions and Prejudices? And what more proper Season to receive those Assistances, and to lay a Foundation for this difficult but very important Science, than the early Part of Youth?

It may be said, " it is properly the Office and Care of Parents, to watch over and correct the Tempers of their Children in the first Years of their Infancy, when it may easiest be done." But if it be not done effectually *then*, (as it very seldom is) there is the more Necessity for it afterwards. But the Truth is, it is the proper Office and Care of all who have the Charge of Youth, and ought to be looked upon as the most important and necessary Part of Education,

It was the Observation of a great Divine and Reformer, that *he who acquires his Learn-*

(d) *Rule of Life, pag.* 82.

ing at the Expence of his Morals is the worse for his Education (e). And we may add, that he who does not improve his Temper, together with his Understanding, is not much the better for it. For he ought to measure his Progress in Science by the Improvement of his Morals; and remember that he is no further a learned Man than he is a wise and good Man; and that he cannot be a finished Philosopher till he is a *Christian (f).*

But whence is it that *moral* Philosophy, which was so carefully cultivated in the antient Academy, should be forced in the modern to give Place to *natural,* that was originally designed to be subservient to it? Which is to exalt the Handmaid into the Place of the Mistress *(g).* This appears not only a preposterous, but a pernicious Method of Institution. For as the Mind takes a Turn of Thought in future Life, suitable to the Tincture it hath received in Youth,

it

(e) Qui proficit in literis et deficit in moribus, non proficit sed deficit. *Oecolampadius.* See *Hist. of Pop. Vol.* ii. *p.* 337.

(f) Te in scientiâ profecisse credas quantum in moribus fueris emendatior; eo usque doctum, in quantum bonum : ita philosophum, ut Christianum. *Præf. ad Nem.*

(g) Things were coming to this Pass so early as *Seneca's* Time; who laments that plain and open *Truth* was turned into a dark and intricate *Science.* " Philosophy (says he) is turned into Philolo-
" gy; and that through the Fault both of Masters and Scholars;
" the one teach to dispute, not to live; and the other come to
" them to mend their Wits, not their Manners.—Whereas Phi-
" losophy is nothing else but a Rule of Life. Quid autem philo-
" sophia, nisi vitæ lex est,"

it will naturally conclude, that there is no Necessity to regard, or at least to lay any Stress upon what was never inculcated upon it as a Matter of Importance *then*. And so will grow up in a Neglect or Disesteem of those Things which are more necessary to make a Person a wise and truly understanding Man, than all those Rudiments of Science he brought with him from the School or College.

It is really a melancholy Thing to see a young Gentleman of shining Parts, and a sweet Disposition, who has gone through the common Course of Academical Studies, come out into the World under an absolute Government of his Passions and Prejudices; which have increased with his Learning, and which, when he comes to be better acquainted with human Life and human Nature, he is soon sensible and ashamed of; but perhaps is never able to conquer as long as he lives, for want of that Assistance which he ought to have received in his Education. For a wrong Education is one of those *three Things* to which it is owing (as an antient Christian and Philosopher justly observes) that so few have the right Government of their Passions (*h*).

I would

(*h*) Εγγινονται δε τα φαυλα παθη τη ψυχη δια τριων τουτων· δια κακης αγωγης, εξ αμαθιας υπο κακοεξιας· μη αχθεντες γαρ καλως εκ παιδων ως δυνασθαι κρατειν των παθων εις την αμελειαν αυτων εμπιμπλομεν.—Bad Passions spring up in the Mind three Ways;

I would not be thought to depreciate any Part of human Literature, but should be glad to see this most useful Branch of Science, the Knowledge of the Heart, the detecting and correcting hurtful Prejudices, and the right Government of the Temper and Passions, in more general Esteem; as necessary at once to form the Gentleman, the Scholar, and the Christian.

And if there be any Thing in this short Treatise which may be helpful to Students, who have a Regard to the right Government of their Minds, whilst they are furnishing them with useful Knowledge, I would particularly recommend it to *their* Perusal.

I have nothing further to add, but to desire the Reader's Excuse for the Freedom with which I have delivered my Sentiments in this Matter, and for detaining him so long from his Subject; which I now leave to his candid and serious Thoughts, and the Blessing of Almighty GOD to make it useful to him.

Ways; *viz.* through a *bad Education*, great *Ignorance*, or a *Disorder* in the *animal Frame*. (1.) From a *bad Education*. For if we have not been taught from our Childhood to govern our *Passions*, with all possible Care, they will soon come to have the Government of us. *Nemes. de Nat. Hom. pag.* 182.

THE CONTENTS.

PART I.

CHAP. I.
                                           Page

THE *Nature and Importance of the Subject.*   22

CHAP. II.
The several Branches of Self-Knowledge. We must know what Sort of Creatures we are, and what we shall be.   32

CHAP. III.
The several Relations in which we stand to GOD, *to* CHRIST, *and our Fellow-Creatures.*   40

CHAP. IV.
We must duly consider the Rank and Station of Life in which Providence hath placed us, and what it is that becomes and adorns it.   55

CHAP. V.
Every Man should be well acquainted with his own Talents and Capacities; and in what Manner they are to be exercised and improved to the greatest Advantage.   59

The CONTENTS.

CHAP. VI.
We must be well acquainted with our Inabilities, and those Things in which we are naturally deficient, as well as those in which we excel. — 62

CHAP. VII.
Concerning the Knowledge of our Constitutional Sins. — 64

CHAP. VIII.
The Knowledge of our most dangerous Temptations necessary to Self-Knowledge. — 71

CHAP. IX.
Self-Knowledge discovers the secret Prejudices of the Heart. — 75

CHAP. X.
The Necessity and Means of knowing our natural Tempers. — 89

CHAP. XI.
Concerning the secret Springs of our Actions. — 94

CHAP. XII.
Every one that knows himself, is in a particular Manner sensible how far he is governed by a Thirst for Applause. — 97

CHAP. XIII.
What Kind of Knowledge we are already furnished with, and what Degree of Esteem we set upon it. — 102

The CONTENTS.

Chap. XIV.
Concerning the Knowledge, Guard, and Government of our Thoughts. ... 108

Chap. XV.
Concerning the Memory. ... 122

Chap. XVI.
Concerning the Mental Taste. ... 126

Chap. XVII.
Of our great and governing Views in Life. ... 131

Chap. XVIII.
How to know the true State of our Souls; and whether we are fit to die. ... 134

PART II.

Shewing the great Excellence and Advantage of this Kind of Science. ... 138

Chap. I.
Self-Knowledge the Spring of Self-possession. ... 139

Chap. II.
Self-Knowledge leads to a wise and steady Conduct. ... 143

Chap. III.
Humility the Effect of Self-Knowledge. ... 145

Chap. IV.
Charity another Effect of Self-Knowledge. ... 148

The CONTENTS.

CHAP. V.
Moderation the Effect of Self-Knowledge. — 151

CHAP. VI.
Self-Knowledge improves the Judgment. — 154

CHAP. VII.
Self-Knowledge directs to the proper Exercises of Self-denial — 156

CHAP. VIII.
Self-Knowledge promotes our Usefulness in the World. — 160

CHAP. IX.
Self-Knowledge leads to a Decorum and Consistency of Character. — 163

CHAP. X.
Piety the Effect of Self-Knowledge. — 166

CHAP. XI.
Self-Knowledge teaches us rightly to perform the Duties of Religion. — 167

CHAP. XII.
Self-Knowledge the best Preparation for Death. — 170

PART III.

Shewing how Self-Knowledge is to be attained. — 177

CHAP. I.
Self-examination necessary to Self-Knowledge. — 177

CHAP. II.
Constant Watchfulness necessary to Self-Knowledge. — 194

CHAP. III.
We should have some Regard to the Opinions of others concerning us, particularly of our Enemies. — 196

CHAP. IV.
Frequent Converse with Superiors a Help to Self-Knowledge. — 201

CHAP. V.
Of cultivating such a Temper as will be the best Disposition to Self-Knowledge. — 202

CHAP. VI.
To be sensible of our false Knowledge a good Step to Self-Knowledge. — 205

CHAP. VII.
Self-inspection especially necessary upon some particular Occasions. — 207

CHAP. VIII.
To know ourselves we must wholly abstract from external Appearances. — 213

CHAP. IX.
The Practice of Self-Knowledge a great Means to promote it. — 215

CHAP. X.
Fervent and frequent Prayer the most effectual Means for attaining true Self-Knowledge. — 223

A TREA-

A TREATISE OF SELF-KNOWLEDGE.

PART I.

CHAP. I.

The Nature and Importance of the Subject.

A DESIRE of Knowledge is natural to the Mind of Man. And nothing difcovers the true Quality and Difpofition of the Mind more, than the particular Kind of Knowledge it is moſt fond of.

Thus we ſee that low and little Minds are moſt delighted with the Knowledge of *Trifles:* as in Children: An indolent Mind, with that which ſerves only for *Amuſement,* or the Entertainment of of the Fancy. A curious Mind is best pleased with *Facts.* A judicious penetrating Mind, with *Demonstration* and Mathematical Science. A worldly Mind esteems no Knowledge like that of the *World.* But a wise and pious Man before all other Kinds of Knowledge prefers that of GOD and his own Soul.

But some Kind of Knowledge or other the Mind is continually craving after. And by considering what that is, its prevailing Turn and Temper may easily be known.

This *Desire of Knowledge,* like other Affections planted in our Nature, will be very apt to lead us wrong, if it be not well regulated. When it is directed to improper Objects, or pursued in a wrong Manner, it degenerates into a vain and criminal *Curiosity.* A fatal Instance of this in our first Parents we have upon Sacred Record; the unhappy Effects of which are but too visible in all.

Self-Knowledge is the Subject of the ensuing Treatise.—A Subject, which the more I think of, the more important and extensive it appears. So important, that every Branch of it seems absolutely necessary to the right Government of the Life and Temper. And so extensive, that the nearer View we take of its several Branches, the more are still opening to the View, as nearly connected with it as the other. Like what we find in *microscopical Observations* on natural Objects. The better the Glasses, and the nearer the Scrutiny, the more Wonders we explore; and the

more

more surprizing Discoveries we make of certain Properties, Parts, or Affections belonging to them, which were never before thought of. For in order to a true *Self-Knowledge*, the Human Mind, with its various Powers and Operations, must be narrowly inspected; all its secret Bendings and Doublings displayed. Otherwise our Self-acquaintance will be but very partial and defective; and the Heart after all will deceive us. So that in treating this Subject there is no small Danger, either of doing Injury to it, by slight and superficial Inquest *on the one Hand*, or of running into a Research too minute and philosophical for common Use *on the other*. The two extremes I shall keep in my Eye, and endeavour to steer a middle Course between them.

Know thyself, is one of the most useful and comprehensive Precepts in the whole moral System. And it is well known in how great a Veneration this Maxim was held by the Antients; and in how high Esteem the Duty of *Self-examination*, as necessary to it.

Thales the Milesian is said to be the first Author of it (*a*). Who used to say, that *for a Man to know himself is the hardest Thing in the World* (*b*). It was afterwards adopted by *Chylon* the *Lacedemonian*; and is one of those three Precepts which

Pliny

(*a*) He was the Prince of the Philosophers, and flourished about *A. M.* 3330. and was contemporary with *Josiah* King of *Judah*.

(*b*) See *Stanley*'s Life of *Thales*.

Pliny affirms to have been confecrated at *Delphos* in golden Letters. It was afterwards greatly admired, and frequently ufed by others (*c*). Till at length, it acquired the Authority of a Divine Oracle; and was fuppofed to have been given originally by *Apollo* himfelf. Of which general Opinion *Cicero* gives us this Reafon; " becaufe " it hath fuch a Weight of Senfe and Wifdom in " it as appears too great to be attributed to any Man (*d*)." And this Opinion, of its coming originally

---

(*c*) Refpue quod non es : tollat fua munera Cerdo.
    Tecum habita: et noris quam fit tibi curta fupellex.
                                                          Perf. Sat. 4.
— — — — nec te quæfiveris extra.   *Id. Sat.* 1.
— — — te confule, dic tibi quis fis.   *Juv. Sat.* 11.
Teipfum concute.   *Hor. lib.* 1. *Sat.* 3.
Bellum eft enim fua vitia noffe.   *Cic. Epift ad Atticum, lib.* 2.

Illud (γνωθι σεαυτον) noli putare ad arrogantiam minuendam folùm effe dictum, verùm etiam ut bona noftra norimus. *Id. Epift. ad Mar. Q. Fratrem, Lib.* 3. *Epift.* 6.

Id enim maximè quemque decet quod eft cujufque fuum maximè. Quifque igitur nofcat Ingenium acremque fe et bonorum et vitiorum fuorum Judicem præbeat. *Id. De Offic. lib.* 1.

Intrandum eft igitur in rerum naturam, et penitus; quid ea poftulat pervidendum; aliter enim nofmet ipfos noffe non poffumus. *Id. De Finibus. lib.* 5.

(*d*) Hæc enim (i. e. Philofophia) nos cùm cæteras res omnes, tum quod eft difficilimum, docuit; ut [NOSMET IPSOS] nofceremus. Cujus Præcepti tanta vis, tanta fententia eft, ut ea non Homini cuipiam, fed Delphico Deo tribueretur. *Cicero De Legib. lib.* 1.

Quod Præceptum quia majus erat quam ut ab Homine videretur, idcirco affignatum eft Deo: Jubet igitur nos Pythius Apollo, nofcere [NOSMET IPSOS.] *Idem De Finibus, lib.* 5. *cap.* 16.

originally from *Apollo* himself, perhaps was the Reason that it was written in golden Capitals over the Door of his Temple at *Delphos*.

And why this excellent Precept should not be held in as high esteem in the *Christian* World as it was in the *Heathen*, is hard to conceive. Human Nature is the same now as it was then. The Heart as deceitful; and the necessity of watching, knowing, and keeping it, the same. Nor are we less assured that this Precept is divine. Nay, we have a much greater Assurance of this than Heathens had; they *supposed* it came down from Heaven, we know it did; what they conjectured, we are sure of. For this sacred Oracle is dictated to us in a manifold Light, and explained to us in various Views by the Holy Spirit, in that Revelation which GOD hath been pleased to give us as our Guide to Duty and Happiness; by which *as in a Glass* we may survey ourselves, and know *what manner of Persons we are* *.

This discovers ourselves to us; pierces into the inmost Recesses of the Mind: strips off every Disguise; lays open the *inward Part*; makes a strict Scrutiny into the very *Soul* and *Spirit*; and *critically judges of the Thoughts and Intents of the Heart* (e). It shows us with what Exactness and Care we are
to

Et nimirum hanc habet vim Præceptum Apollinis, quo monet ut se quisque noscat—Hunc igitur nosse, (i. e. animum) nisi divinum esset, non esset hoc acrioris cujusdam animi Preceptum, sic, ut tributum Deo sit: hoc est seipsum posse cognoscere. *Idem Tuscul. Quæst. lib.* 5.

* Jame. i. 23.

(e) Και κριτικος ενθυμησεων και εννοιων καρδιας. *Heb.* iv. 12.

to search and try our Spirits, examine ourselves, and watch our Ways, and keep our Hearts, in order to acquire this important Self-science; which it often calls us to do. *Examine yourselves,* — *Prove your ownselves*; *Know you not yourselves,(f)*? *Let a Man examine himself**. Our Saviour upbraids his Disciples with their Self-ignorance, in not knowing *what Manner of Spirits they were of* †. And saith the Apostle, *If a Man* (through Self-ignorance) *thinketh himself to be Something, when he is Nothing, he deceiveth himself. But let every Man prove his Work, and then shall he have rejoicing in himself, and not in another* ‡. Here we are commanded, instead of *judging* others, to *judge* ourselves; and to avoid the *inexcusable Rashness* of condemning others for the very Crimes we ourselves are guilty of, *Rom.* ii. 1, 21, 22. which a Self-ignorant Man is very apt to do; nay, to be more offended at a small Blemish in another's Character, than at a greater

(f) Ἑαυτοὺς δοκιμάζετε. 2 Cor. xiii. 5.—Tho' δοκιμάζειν signifies to *approve* as well as to *prove*, yet that our translators have hit upon the true Sense of the Word here, in rendering it *prove yourselves*, is apparent, not only from the Word immediately preceding (ἑαυτοὺς πειράζετε) which is of the same Import, but because *Self-probation* is always necessary to a right *Self-approbation*.

" Every Christian ought to try himself, and may know him-
" self if he be faithful in examining. The frequent Exhortations
" of Scripture hereunto imply both these, *viz.* that the Know-
" ledge of ourselves is attainable, and that we should endeavour
" after it. Why should the Apostle put them upon examining
" and proving themselves, unless it was possible to know them-
" selves upon such trying and proving?" *Bennet's Christ. Gratory, p.* 568.

* 1 Cor xi. 28.   † *Luke* ix. 55,   ‡ *Gal* vi. 3. 4.

in his own; which Folly, Self-ignorance, and Hypocrisy, our Saviour with just Severity animadverts upon, *Matt.* vii. 3,—5.

And what Stress was laid upon this under the Old Testament Dispensation appears sufficiently from those Expressions. *Keep thy Heart with all Diligence* *. *Commune with your own Heart* †. *Search me, O GOD, and know my Heart; try me, and know my Thoughts* ‡. *Examine me, O LORD, and prove me; try my Reins and my Heart* ‖. *Let us search and try our Ways* §. *Recollect, recollect yourselves, O Nation not desired* ** (*g*).—And all this as necessary to that *Self-acquaintance* which is the only proper Basis of solid Peace (*h*).

Were mankind but more generally convinced of the Importance and Necessity of this *Self-Knowledge*, and possessed with a due Esteem for it;

did

---

* *Prov.* iv. 23.   † *Psal.* iv. 4.   ‡ *Psal.* cxxxix. 23.
‖ *Psal.* xxvi. 2.   § *Lam.* iii. 4.   ** *Zeph.* ii. 1.

(*g*) התקוששו וקושו—the Verb (קשש) properly signifies to *glean*, or gather together scatter'd Sticks or Straws; as appears from all the Places where the Word is used in the Old Testament (*Exod.* v. 7, 12. *Num.* xv. 32. 1 *Kings* xvii. 10.) Hence by an easy Metaphor it signifies to *recollect*, or gather the scattered Thoughts together; and ought to be so rendered, when used in the reflective Form, as here it is. So saith R. Kimchi, (קשש) est propriè stipulas colligere. Id fit accuratâ scrutatione hinc d'citur de qualibet Inquisitione. Whence I think it is evident that the Word should be rendered as above.

(*b*) *Clement Alexandrinus* saith, that *Moses* by that Phrase, so common in his Writings, *Take heed to thyself* (*Exod.* x. 28. xxxiv. 12. *Deut.* iv. 9.) means the same Thing as the Antients did by their γνῶθι σεαυτὸν. *Strom.* lib. 2. cap. 5.

did they but know the true Way to attain it; and under a proper Sense of its Excellence, and the fatal Effects of Self-ignorance, did they but make it their Business and Study every Day to cultivate it; how soon should we find a happy Alteration in the Manners and Spirits of Men!——But the Misery of it is, Men will not *think*; will not employ their Thoughts, in good Earnest, about the Things which most of all deserve and demand them. By which unaccountable Indolence, and Aversion to Self-reflection, they are led blindfold and insensibly into the most dangerous Paths of Infidelity and Wickedness, as the *Jews* were heretofore; of whose amazing Ingratitude and Apostacy GOD himself assigns this single Cause? * *My People do not consider* (i).

Self-Knowledge is that Acquaintance with ourselves, which shews us what we are, and do, and ought to be, and do, in order to our living comfortably and usefully here, and happily hereafter. The Means of it is *Self-examination*; the End of it *Self-government*, and *Self-fruition.*—— It principally consists in the Knowledge of our *Souls*; which is attained by a particular Attention to their various Powers, Capacities, Passions, Inclinations, Operations, State,

Happi-

---

* *Isai.* i. 3.

(i) "There is nothing Men are more deficient in, than "knowing their own Characters. I know not how this Sci- "ence comes to be so much neglected. We spend a great deal of "Time in learning useless Things, but take no Pains in the "Study of ourselves; and in opening the Folds and Doubles of "the Heart." *Reflections on Ridicule, Pag.* 61.

Happiness, and Temper. For a Man's Soul is properly himself, *Mat.* xvi. 26. compared with *Luke* ix. 25. (*k*). The Body is but the House, the Soul is the Tenant that inhabits it; the Body is the Instrument, the Soul the Artist that directs it (*l*).

This Science, which is to be the Subject of the ensuing Treatise, hath these three peculiar Properties in it, which distinguish it from, and render it preferable to all other. — (1.) *It is equally attainable by all.* It requires no Strength of Memory, no Force of Genius, no Depth of Penetration, as many other Sciences do, to come at a tolerable Degree of Acquaintance with them; which therefore renders them *inaccessible* by the greatest Part of

---

(*k*) Præceptum Apollinis quo monet, ut se quisque noscat, non enim, credo, id præcipit; ut Membra nostra aut Staturam Figuramque noscamus: neque nos corpora sumus; neque ego, tibi disens hoc, Corpori tuo dico: cum igitur NOSCE TE dicit, hoc dicit, Nosce animum tuum. Nam Corpus quidem quasi vas est, aut aliquod Animi Receptaculum; ab Animo tuo quicquid agitur id agitur a te. *Cic. Tuscul. Quæst. lib.* 1.

(*l*) 2 *Cor.* v. 1. *Rom.* vi. 13. —— ἡ δυναμις ψυχης, το δε οργανον σωυματος. *Nemes. de Nat. Hom. cap.* 6.

Μηδεποτε σμπεριφανταξη το περικειμενον αγγειωδες και τα οργανια ταυτα τα περιπεπλασμενα, ομοια γαρ εστι σκιταγρω, μονον δε διαφεροντα, καθοτι προσφυη εστιν. *Mar. Anton. lib.* x. § 37. When you talk of a Man, I would not have you tack Flesh and Blood to the Notion, nor those Limbs neither which are made out of it; these are but Tools for the Soul to work with: and no more a Part of a Man, than an Axe or a Plane is a Piece of a Carpenter. It is true, Nature hath glewed them together, and they grow as it were to the Soul, and there is all the Difference. *Collier*.

of Mankind. Nor is it plac'd out of their Reach thro' a Want of Opportunity, and proper Affiftance and Direction how to acquire it; as many other Parts of Learning are. Every one of a common Capacity hath the Opportunity and Ability to attain it, if he will but recollect his rambling Thoughts, turn them in upon himfelf, watch the Motions of his Heart, and compare them with his Rule. — (2.) *It is of equal Importance to all; and of the higheft Importance to every one* (*m*). Other Sciences are fuited to the various Conditions of Life. Some, more neceffary to fome; other, to others. But this equally concerns every one that hath an immortal Soul, whofe final Happinefs he defires and seeks. —— (3.) *Other Knowledge is very apt to make a Man vain; this always keeps him humble.* Nay, it is for want of *this* Knowledge that Men are vain of that they have. *Knowledge puffeth up* †. A fmall Degree of Knowledge often hath this Effect on weak Minds. And the Reafon why greater Attainments in it have not fo generally the fame Effect is, becaufe they open and enlarge the Views of the Mind fo far, as to let into it at the fame Time a good Degree of *Self-Knowledge.* For the more true Knowledge a Man hath, the more fenfible he is of the Want of it; which keeps him humble.

And

(*m*) 'Tis Virtue only makes our Blifs below,
  And all our Knowledge is OURSELVES TO KNOW.
                                       Pope's Effay on Man.

† 1 *Cor.* viii. 1.

And now, *Reader*, whoever thou art, whatever be thy Character, Station, or Distinction in Life, if thou art afraid to look into thine Heart, and hast no Inclination to Self-acquaintance, read no further: lay aside this Book; for thou wilt find nothing here that will flatter thy Self-esteem; but perhaps something that may abate it. But if thou art desirous to cultivate this important Kind of Knowledge, and to live no longer a Stranger to thyself, proceed; and keep thy Eye open to thine own Image, with whatever unexpected Deformity it may present itself to thee; and patiently attend, whilst, by Divine Assistance, I endeavour to lay open thine own Heart to thee, and lead thee to the true Knowledge of thyself in the following Chapters.

CHAP. II.

The several Branches of Self-Knowledge. We must know what Sort of Creatures we are, and what we shall be.

THAT we may have a more distinct and orderly View of this Subject, I shall here consider the several Branches of Self-Knowledge; or some of the chief Particulars wherein it consists. Whereby perhaps it will appear to be a more copious and comprehensive Science than we imagine. And,

(1.) To

(1.) To know ourselves, is *to know and seriously consider what Sort of Creatures we are, and what we shall be.*

(1.) *What we are.*

Man is a complex Being, τριμερης υποστασις, a *tripartite Person*; or a compound Creature made up of three distinct Parts, *viz.* the *Body*, which is the earthly or mortal Part of him, the *Soul*, which is the animal or sensitive Part: and the *Spirit* or *Mind*, which is the rational and immortal Part *.—Each of these three Parts have their

---

* This Doctrine, I think, is established beyond all Dispute, not only by Experience, but by Authority. It was received by almost all the antient Philosophers. The *Pythagoreans*; as we learn from *Jamblicus, vid.* Protrept. *p.* 34, 35. The *Platonists*; as appears from *Nemesius, Sallust,* and *Laertius, vid. Di. Laertius, lib.* 3. *p.* 219. The *Stoicks*; as appears from *Antoninus,* who saith expresly, "There are three Things which belong to a Man; the "Body, Soul, and the Mind. And as to the Properties of the "Division, Sensation belongs to the Body, Appetite to the Soul, "and Reason to the Mind," σωμα, ψυχη, νυς, σωματος αισθησεις, ψυχης ορμαι, νε δογματα. *lib.* 3. § 16. *lib.* 2. § 2. *lib.* 12. § 3. ——It appears also to have been the Opinion of most of the Fathers, *vid. Irenæus, lib.* 5. *cap.* 9. *lib.* 2. *cap.* 33. *Ed. Par. Clem. Alex. Strom.* 3. *p.* 542. *Ed. Oxon. Origen. Philocal. p.* 8. *Ignat. Ep. ad Philadelph. ad calcem.* See also *Joseph. Antiq. lib.* 1. *cap.* 2. *p.* 5. *Constitut. Apostol. lib.* 7. *cap.* 34.——But above all these, is the Authority of Scripture, which speaking of the original Formation of Man, mentions the three distinct Parts of his Nature; *Gen.* ii. 7. *viz.* עָפָר מִן־הָאֲדָמָה *the Dust of the Earth*, or the *Body;* נֶפֶשׁ חַיָּה *the living Soul*, or the animal and sensitive Part: and נִשְׁמַת חַיִּים *the Breath of Life,* i. e. the Spirit or rational Mind. In like Manner the Apostle *Paul* divides the *whole Man* into (το πνευμα, η ψυχη, και το σωμα) the *Spirit*, the *Soul*, and the

their respective Offices assigned them. And a Man then acts becoming himself, when he keeps them duly employed in their proper Functions, and preserves their natural Subordination.—But it is not enough to know this merely as a Point of Speculation; we must pursue and revolve the Thought, and urge the Consideration to all the Purposes of a practical Self-acquaintance.

We are not all Body, nor mere animal Creatures. We find we have a more noble Nature than the inanimate, or brutal Part of the Creation. We can not only move and act freely, but we observe in ourselves a Capacity of Reflection, Study, and Forecast; and various mental Operations, which irrational Animals discover no Symptoms of. *Our* Souls therefore must be of a more excellent Nature than *theirs*; and from the Power

---

the *Body*, 1 Thess. v. 23. and what he calls (πνεῦμα) here, he calls (νᾶς) Rom. vii. 24. the Word which *Antoninus* uses to denote the same Thing.—They who would see more of this may consult *Nemesius de Naturâ Hominis*, cap. 1. and *Whiston's Prim. Christ. vol.* 4. *pag.* 262.

All the Observation I shall make hereupon is, that this Consideration may serve to soften the Prejudices of some against the Account which Scripture gives us of the mysterious Manner of the Existence of the divine Nature: of which every Man (as *created in the Image of GOD*) carries about him a kind of Emblem, in the threefold Distinction of his own; which, if he did not every Minute find it by Experience to be Fact, would doubtless appear to him altogether as mysterious and incomprehensible as the *Scripture-doctrine* of the Trinity.

" Homo habet tres Partes, Spiritum, Animam, et Corpus; " itaque Homo est Imago S. S. Trinitatis." *August. Tractat. de Symbolo.*

Power of Thought with which they are endowed, they are proved to be immaterial Substances: And consequently in their own Nature capable of Immortality. And that they are actually immortal, or will never die, the sacred Scriptures do abundantly testify (*m*).—Let us then hereupon seriously recollect ourselves in the following Soliloquy.

'O my Soul, look back but a few Years, and
'thou wast nothing!——And how didst thou
'spring out of that Nothing?—Thou couldst not
'make thyself. That is quite impossible.—Most
'certain it is, that *that* Almighty, self-existent
'and eternal Power, which made the World,
'made thee also out of nothing. Called thee into
'Being when thou wast not; gave thee these
'reasoning and reflecting Faculties, which thou
'art now employing in searching out the End and
'Happiness of thy Nature.—It was He, O my
'Soul, that made thee intelligent and immortal.
'It was He that placed thee in this Body, as in a
'Prison;

(*m*) As Nature delights in the most easy Transitions from one Class of Beings to another, and as the *Nexus utriusque Generis* is observable in several Creatures of ambiguous Nature, which seem to connect the lifeless and vegetable, the vegetable and animal, the animal and rational Worlds together; (See *Nemesius de Nat. Hom. cap. 1. p. 6.*) Why may not the Souls of Brutes be considered as the *Nexus* between material and immaterial Substances, or Matter and Spirit, or *something between* both? The great Dissimilitude of Nature in these two Substances, I apprehend, can be no solid Objection to this *Hypothesis*, if we consider (beside our own Ignorance of the Nature of Spirits) but how nearly they approach in other Instances, and how closely they are *united* in Man.

'Prison; where thy Capacities are cramped, thy
'Desires debased, and thy Liberty lost.—It was
'He that sent thee into this World, which by all
'Circumstances appears to be a State of short
'Discipline and Trial. And wherefore did He
'place thee here, when he might have made thee
'a more free, unconfined, and happy Spirit?—
'But check that Thought;—it looks like a too
'presumptuous Curiosity. A more needful and
'important Enquiry is; what did He place thee
'here for? And what doth He expect from thee,
'whilst thou art here?—What Part hath he al-
'lotted me to act on the Stage of human Life;
'where He, Angels and Men, are Spectators of
'my Behaviour? The Part He hath given me to
'act here is, doubtless, a very important one;
'because it is for Eternity (*n*). And what is it,
'but to live up to the Dignity of my rational
'and intellectual Nature; and as becomes a
'Creature born for Immortality?

'And tell me, O my Soul, (for as I am now
'about to cultivate a better Acquaintance with
'thee, to whom I have been too long a Stranger,
'I must try thee, and put many a close Question
'to thee,) tell me, I say, whilst thou confinest
'thy Desires to sensual Gratifications, wherein
'dost

(*n*) It is said when the Prince of the Latin Poets was asked by his Friend, why he studied so much Accuracy in the Plan of his Poem, the Propriety of his Characters, and the Purity of his Diction; he replied, In æternum pingo, *I am writing for Eternity*. What more weighty Consideration to justify and inforce the utmost Vigilance and Circumspection of Life, than this; In æternum vivo, *I am living for Eternity?*

'doſt thou differ from the Beaſts *that periſh?*
'Captivated by bodily Appetites, doſt thou not
'act beneath thyſelf? Doſt thou not put thyſelf
'upon a Level with the lower Claſs of Beings,
'which were made to ſerve thee, offer an Indig-
'nity to thyſelf, and deſpiſe the Work of thy
'Maker's Hands? O remember thy heavenly
'Extract; remember thou art a Spirit. Check
'then the Solicitations of the Fleſh; and dare to
'do nothing that may diminiſh thy native Ex-
'cellence, diſhonour thy high Original, or de-
'grade thy noble Nature (*o*).— But let me ſtill
'urge it. Conſider, I ſay, O my Soul, that thou
'art an immortal Spirit. Thy Body dies; but
'thou, *thou* muſt live for ever, and thine Eter-
'nity will take its Tincture from the Manner of
'thy Behaviour, and the Habits thou contracteſt,
'during this thy ſhort Co-partnerſhip with Fleſh
'and Blood. O! do nothing now, but what
'thou mayeſt with Pleaſure look back upon a
'Million of Ages hence. For know, O my Soul,
'that thy Self-conſciouſneſs, and reflecting Fa-
'culties will not leave thee with thy Body; but
'will follow thee after Death, and be the In-
'ſtrument

(*o*) Major ſum et ad majora natus, quam quod ſim corporis mancipium. Quod equidem non aliter aſpicio quam vinculum libertati meæ circumdatum. *Sen. Ep.* 66.

I am too noble, and of too high a Birth (ſaith that excellent Moraliſt) *to be a Slave to my Body; which I look upon only as a Chain thrown upon the Liberty of my Soul.*

'ſtrument of unſpeakable Pleaſure or Torment
' to thee in that ſeparate State of Exiſtence*.'

(2.) *In order to a full Acquaintance with our-ſelves, we muſt endeavour to know not only what we are, but what we ſhall be.*

And O! what different Creatures ſhall we *ſoon* be, from what we *now* are! Let us look forwards then, and frequently glance our Thoughts towards Death; tho' they cannot penetrate the Darkneſs of that Paſſage, or reach the State behind it. *That* lies veil'd from the Eyes of our Mind. And the great GOD hath not thought fit to throw ſo much Light upon it, as to ſatisfy the anxious and inquiſitive Deſires the Soul hath to know it. However, let us make the beſt Uſe we can of that little Light which Scripture and Reaſon have let in upon this dark and important Subject.

' Compoſe thy Thoughts, O my Soul, and i-
' magine how it will fare with thee, when thou
' goeſt a naked, unimbodied Spirit, into a World,
' an unknown World of Spirits, with all thy
' Self-conſciouſneſs about thee, where no materi-
' al Object ſhall ſtrike thine Eye; and where thy
' dear

---

* As it is not the Deſign of this Treatiſe to enter into a nice and philoſophical Diſquiſition concerning the Nature of the human Soul, but to awaken Men's Attention to the inward Operations and Affections of it, (which is by far the moſt neceſſary Part of Self-Knowledge) ſo they who would be more paticularly informed concerning its Nature and Original, and the various Opinions of the Antients about it, may conſult *Nemeſ. de Nat. Hom. cap.* 1. and a Treatiſe called *The Government of the Thoughts, chap.* 1, and *Chambers's Cyclopædia*, under the Word SOUL.

'dear Partner and Companion the Body cannot
'come nigh thee. But where without it thou
'wilt be sensible of the most noble Satisfactions,
'or the most exquisite Pains. Imbarked in Death,
'thy Passage will be dark; and the Shore, on
'which it will land thee, altogether strange and
'unknown.—— *It doth not yet appear what we
'shall be* (p).'

That Revelation, which GOD hath been pleased to make of his Will to Mankind, was designed

(p) 'Thou must expire, my Soul, ordain'd to range
'Thro' unexperienc'd Scenes, and Myst'ries strange;
'Dark the Event, and dismal the Exchange.
'But when compell'd to leave this House of Clay,
'And to an unknown *Somewhere* wing thy Way;
'When Time shall be Eternity, and Thou
'Shalt be thou know'st not what, nor where, nor how,
'Trembling and pale, what wilt thou see or do?
'Amazing State!——No wonder that we dread
'The Thoughts of Death, or Faces of the Dead.
'His black Retinue sorely strikes our Mind;
'Sickness and Pain before, and Darkness all behind.
  'Some courteous Ghost, the Secret then reveal;
'Tell us what you have felt, and we must feel.
'You warn us of approaching Death, and why
'Will you not teach us what it is to die?
'But having shot the Gulph, you love to view
'Succeeding Spirits plung'd along like you;
'Nor lend a friendly Hand to guide them through.
  'When dire Disease shall cut, or Age untie
'The Knot of Life, and suffer us to die:
'When after some Delay, some trembling Strife,
'The Soul stands quiv'ring on the Ridge of Life;
'With Fear and Hope she throbs, then curious tries
'Some strange *Hereafter*, and some hidden Skies.'

Norris.

signed rather to fit us for the future Happiness, and direct our Way to it, than open to us the particular Glories of it; or distinctly shew us what it is. This it hath left still very much a Mystery; to check our too curious Enquiries into the Nature of it, and to bend our Thoughts more intently to that which more concerns us, *viz.* an habitual Preparation for it. And what that is, we cannot be ignorant, if we believe either our Bible or our Reason. For both these assure us, that *that which makes us like to GOD, is the only Thing that can fit us for the Enjoyment of Him.*—Here then let us hold. Let our great Concern be, to be *holy as he is holy.* And then, then only, are we sure to enjoy him, *in whose Light we shall see Light.* And be the future State of Existence what it will, we shall some Way be happy there. And much more happy than we can now conceive; tho' in what particular Manner we know not, because GOD hath not revealed it.

CHAP. III.

The several Relations wherein we stand to GOD, to CHRIST, and our Fellow-Creatures.

II. SELF-KNOWLEDGE *requires us to be well acquainted with the various Relations in which we stand to other Beings, and the several*

several Duties that result from those Relations. And,

(1.) *Our first and principal Concern is to consider the Relation wherein we stand to Him who gave us Being.*

We are the Creatures of his Hand, and the Objects of his Care. His Power up holds the Being his Goodness gave us. His Bounty accommodates us with the Blessings of *this* Life, and his Grace provides for us the Happiness of a better. —— Nor are we merely his Creatures, but his rational and intelligent Creatures. It is the Dignity of our Natures, that we are capable of knowing and enjoying him that made us. And as the rational Creatures of GOD, there are two Relations especially that we bear to Him; the frequent Consideration of which is absolutely necessary to a right *Self-Knowledge.* For as our Creator, He is our *King* and *Father.* And as his Creatures, we are the *Subjects* of his Kingdom, and the *Children* of his Family.

(1.) *We are the Subjects of his Kingdom.* And as such we are bound,

(1.) To yield a faithful *Obedience* to the Laws of his Kingdom. —— And the Advantages by which these come recommended to us above all Human Laws are many. — They are calculated for the *private* Interest of every one, as well as that of the Public; and are designed to promote our present, as well as our future Happiness. — They are plainly and explicitly published; easily
under-

understood; and in fair and legible Characters writ in every Man's Heart; and the Wisdom, Reason, and Necessity of them are readily discerned.—They are urged with the most mighty Motives that can possibly affect the human Heart.—And if any of them are difficult, the most effectual Grace is freely offered to encourage and assist our Obedience: advantages which no human Laws have to enforce the Observance of them. — (2.) As his *Subjects* we must readily pay him the *Homage* due to his *Sovereignty*. And this is no less than the Homage of the Heart; humbly acknowledging that we hold every Thing of him, and have every Thing from him. Earthly Princes are forced to be content with verbal Acknowledgments, or mere formal Homage. For they can command nothing but what is external. But GOD, who knows and looks at the Hearts of all his Creatures, will accept of nothing but what comes from thence. He demands the Adoration of our whole Souls, which is most justly due to him who formed them, and gave them the very Capacities to know and adore him.—(3.) As faithful *Subjects*, we must chearfully pay him the *Tribute* he requires of us. This is not like the Tribute which earthly Kings exact; who as much depend upon their Subjects for the Support of their Power, as their Subjects do upon them for the Protection of their Property. But the Tribute GOD requires of us, is a Tribute of Praise and Honour, which he stands in no need of from us.

us. For his Power is independent, and his Glory immutable; and he is infinitely able of himself to support the Dignity of his universal Government. But it is the most *natural* Duty we owe to him as Creatures. For to praise him, is only to shew forth his Praise; to glorify him, to celebrate his Glory; and to honour him, is to render him and his Ways honourable in the Eyes and Esteem of others. And as this is the most natural Duty that Creatures owe to their Creator, so it is a Tribute he requires of every one of them in Proportion to their respective Talents and Abilities to pay it.—— (4.) As dutiful *Subjects*, we must contentedly and quietly submit to the *Methods* and *Administrations* of his Government, however dark, involved or intricate. All Governments have their *Arcana Imperii,* or *Secrets of State;* which common Subjects cannot penetrate. And therefore they cannot competently judge of the Wisdom or Rectitude of certain public Measures; because they are ignorant eitheir of the Springs of them, or the Ends of them, or the Expediency of the Means arising from the particular Situation of Things in the present Juncture. And how much truer is this with relation to GOD's Government of the World? whose Wisdom is far above our Reach, and *whose Ways are not as ours.* Whatever then may be the present Aspect and Appearance of Things, as dutiful Subjects we are bound to acquiesce; to ascribe Wisdom and *Righteousness to our Maker,* in Confidence that

the

the King and *Judge of all the Earth will do right.*
——Again, (5.) As good Subjects of GOD's Kingdom, we are bound to pay a due Regard and Reverence to his *Ministers*; especially if they discover an uncorrupted Fidelity to his Cause, and a pure unaffected Zeal for his Honour; if they do not seek their own Interest more than that of their divine Master. The Ministers of earthly Princes too often do this, and it would be happy if all the Ministers and Ambassadors of the heavenly King were intirely clear of the Imputation.—It is no uncommon Thing for the Honour of an earthly Monarch to be wounded thro' the Sides of his Ministers. The Defamation and Slander that is directly thrown at *them*, is obliquely intended against *Him*; and as such it is taken. So to attempt to make the Ministers of the Gospel, in general, the Objects of Derision, as some do, plainly shews a Mind very dissolute and disaffected to GOD and Religion itself; and is to act a Part very unbecoming the dutiful Subjects of his Kingdom.—(Lastly,) As good Subjects, we are to do all we can to promote the *Interest* of his Kingdom; by defending the Wisdom of his Administrations, and endeavouring to reconcile others thereunto, under all the Darkness and Difficulties that may appear therein, in Opposition to the profane Censures of the prosperous Wicked, and the Doubts and Dismays of the afflicted Righteous. — This is to act in Character as loyal *Subjects* of the King of Heaven. And whoever

whoever forgets this Part of his Character, or acts contrary to it, shews a great Degree of *Self-Ignorance*.

But, (2.) As the *Creatures* of GOD, we are not only the *Subjects of his Kingdom*, but the *Children of his Family*. And to this Relation, and the Obligations of it, must we carefully attend, if we would attain the true Knowledge of ourselves.—We are his Children by *Creation*; in which respect he is truly our Father. * *But now, O LORD, thou art our Father: we are the Clay, and thou our Potter; and we all are the Work of thine Hands.* And in a more special Sense we are his Children by *Adoption*. † *For ye are all the Children of GOD by Faith in Christ Jesus.*— And therefore, (1.) We are under the highest Obligations to *love* Him as our Father. The Love of Children to Parents is founded on Gratitude for Benefits received, which can never be requited; and ought in Reason to be proportioned to those Benefits. And what Duty more natural than to love our Benefactors? What Love and Gratitude then is due to Him, from whom we have received the greatest Benefit, even that of our Being, and every Thing that contributes to the Comfort of it?—(2.) As his Children, we must *honour* Him; that is, must speak honourably of Him, and for Him; and carefully avoid every Thing that may tend to dishonour his holy Name and Ways. ‡ *A Son honoureth*

---

* *Isai.* lxiv. 8. † *Gal.* iii. 26. ‡ *Mal.* i. 6.

eth his Father:—*if then I be a Father, where is mine Honour?*—(3.) As our Father we are to *apply to* him for what we want. Whither should Children go, but to their Father, for Protection, Help and Relief in every Danger, Difficulty, and Distress?— And (4.) We must *trust* his Power and Wisdom, and paternal Goodness, to provide for us, take care of us, and do for us that which is best; and what that is he knows best. To be anxiously fearful what will become of us, and discontented and perplexed under the Apprehension of future Evils, whilst we are in the Hands and under the Care of our Father which is in Heaven, is not to act like *Children*. Earthly Parents cannot avert from their Children all the Calamities they fear, because their Wisdom and Power are limited; but our All-wise and Almighty Father in Heaven can. They may possibly want Love and Tenderness, but our heavenly Father cannot, *Isai.* xlix. 15.—(5.) As Children, we must quietly *acquiesce* in his Disposals, and not expect to see into the Wisdom of all his Will. It would be indecent and undutiful in a Child to dispute the Authority, or question the Wisdom, or neglect the Orders of his Parents every Time he could not discern the Reason and Design thereof. Much more unreasonable and unbecoming is such a Behaviour towards GOD, *who giveth not Account of any of his Matters; whose Judgments are unsearchable, and whose Ways are past finding out* *.—(Lastly,) As Children, we must patiently

submit

---

* *Job* xxxiii. 13. *Rom.* xi. 33.

submit to his *Discipline* and *Correction*. Earthly Parents may sometimes punish their Children through Passion, or *for their Pleasure*; but our heavenly Father always corrects *his* for their *Profit* *, and only if *need* be †, and never so much *as their Iniquities deserve* ‡. — Under his fatherly Rebukes then let us be ever humble and submissive.—Such now is the true filial Disposition. Such a Temper, and such a Behaviour should we shew towards GOD, if we would act in Character as his *Children*.

These then are the two special Relations, which, as Creatures, we stand in to GOD. And not to act towards him in the Manner beforementioned, is to shew that we are ignorant of, or have not yet duly considered our Obligations to Him as his *Subjects* and his *Children*; or that we are as yet ignorant both of *GOD* and *ourselves*.—Thus we see how directly the Knowledge of ourselves leads us to the Knowledge of GOD. So true is the Observation of a late pious and very worthy Divine, that " He that is a " Stranger to himself, is a Stranger to GOD, " and to every Thing that may denominate " him wise and happy (*q*)."

But, (2.) In order to *know ourselves* there is another important Relation we should often think of, and that is, *That in which we stand to JESUS CHRIST our Redeemer.*

The

* *Heb.* xii. 10.   † 1 *Pet.* i. 6.   ‡ *Ezra* ix. 13.
(*q*) See Mr. *Baxter*'s Dedicatory Epistle, prefix'd to his Treatise on *the Benefits of Self-acquaintance.*

The former was common to us as Men; this is peculiar to us as Christians, and opens to us a new Scene of Duties and Obligations, which a Man can never forget, that does not grosly forget himself. For as Christians, we are the Disciples, the Followers, and the Servants of Christ, redeemed by him.

And, (1.) As the *Disciples* of Christ, we are to *learn* of Him. To take our religious Sentiments only from his Gospel, in Opposition to all the authoritative Dictates of Men, who are weak and fallible as ourselves. *Call no Man Master on Earth.* Whilst some affect to distinguish themselves by Party-Names, as the *Corinthians* formerly did, (for which the Apostle blames them) one saying, *I am of Paul*; another, *I am of Apollos*; another, *I am of Cephas* *, let us remember that we are the Disciples of Christ; and in this Sense *make mention of his Name only*. It is really injurious to it, to seek to distinguish ourselves by any other. There is more Carnality in such Party-distinctions, Denominations and Attachments, than many good Souls are aware of; tho' not more than the Apostle *Paul*; (who was unwillingly placed at the Head of one himself) hath apprised them of †. —— We are *of Christ*; our Concern is, to honour that superior Denomination, by living up to it. And to adhere inflexibly to his Gospel, as the only Rule of our Faith, the Guide of our Life, and the Foundation of our Hope; whatever Contempt or Abuse

---

* 1 *Cor.* i. 12. † 1 *Cor.* iii. 4.

Abuse we may suffer either from the profane or bigotted Part of Mankind for so doing.—(2.) As Christians we are *Followers of* CHRIST; and therefore bound to *imitate* him, and copy after that most excellent Pattern he hath set us, *who hath left us an Example that we should follow his Steps* *. To see that the *same* holy Temper *be in us which was in him*; and to discover it in the same Manner he did, and upon like Occasions. To this he calls us ‡, and no Man is any further a Christian than as he is a Follower of Christ; aiming at a more perfect Conformity to that most perfect Example which he hath set us of universal Goodness.—(3) As Christians, we are the *Servants* of CHRIST; and the various Duties which Servants owe to their Masters in any Degree, those we owe to him in the highest Degree; who expects we should behave ourselves in his Service with that Fidelity and Zeal, and steady Regard to his Honour and Interest, at all Times, which we are bound to by Virtue of this Relation, and to which his unmerited and unlimited Goodness and Love lays us under infinite Obligations.—(Lastly,) We are moreover his *redeemed* Servants; and as such are under the strongest Motives to *love* and *trust* him.

This deserves to be more particularly considered, because it opens to us another View of the human Nature, in which we should often survey ourselves, if we desire to know ourselves; and that is, as *depraved* or *degenerate* Beings. The inward

* 1 *Pet.* ii, 21. ‡ *Mat.* xi. 29.

ward Contest we so sensibly feel, at some Seasons especially, between a good and a bad Principle, (called in Scripture-Language the *Flesh* and the *Spirit*,) of which some of the wisest Heathens seemed not to be ignorant *: this, I say, is Demonstration that some Way or other the Human Nature has contracted an ill Bias, (and how that came about the sacred Scriptures have sufficiently informed us,) and that it is not what it was when it came originally out of the Hands of its Maker; so that the Words which St. *Paul* spake with reference to the Jews in particular, are justly applicable to the present State of Mankind in general, *there is none righteous, no not one;—they are all gone out of the Way, they are together become unprofitable, there is none that doth good, no not one* ‡.

This is a very mortifying Thought; but an undeniable Truth, and one of the first Principles of that Science we are treating of, and very necessary to be attended to, if we would be sensible of the Duty and Obligations we owe to Christ as the great REDEEMER; in which Character he appears for the Relief and Recovery of Mankind under this their universal Depravity.

The two miserable Effects of the Human Apostacy are, (1.) That perverse Dispositions grow up in our Minds from early Infancy, soon

---

* Αυγεη γαρ ανω παδες ερις βλαπτεσα λεληθεν Συμφυτος. *Pythag. Aur. Carm.*

A fatal inbred Strife does lurk within,
The Cause of all this Misery and Sin.

‡ *Rom.* iii. 10, 12.

settle into vicious Habits, and render us weak and unwilling to obey the Dictates of Conscience and Reason: this is commonly called the *Dominion* of Sin. And, (2.) At the same Time we are subject to the Displeasure of GOD, and the Penalty of his Law; which is commonly called the *Condemnation* of Sin. Now in both these Respects did Christ *the Lamb of GOD come to take away the Sin of the World*; that is, to take away the *reigning* Power of it by the Operation of his Grace; and its *condemning* Power by the Atonement of his Blood; to sanctify us by his Spirit, and justify us by his Death; by the former he reconciles us to GOD, and by the latter he reconciles GOD to us (r), and is at once our *Righteousness* and *Strength*. He died to purchase for us the Happiness we had forfeited, and sends his Grace and Spirit to fit us for that Happiness he hath thus purchased. So complete is his Redemption! so precisely adapted is the Remedy he hath provided, to the Malady we had contracted!———' O Blessed Re-
' deemer of wretched ruined Creatures, how un-
' speakable are the Obligations I owe thee! But
' Ah! How insensible am I to those Obligati-
' ons! the saddest Symptom of Degeneracy I find
' in my Nature, is that base Ingratitude of Heart
' which

(r) By this Phrase I do not mean that GOD was implacable or absolutely irreconcileable to us till he was pacified by the vicarious Sufferings of his Son; for how then could he have appointed him to die as our propitiatory Sacrifice? But that the Death of Christ is the clearest Demonstration of GOD's Willingness to be actually reconciled to us.

'which renders me so unaffected with thine a-
'stonishing Compassions. Till I know thee, I
'cannot know myself: and when I survey my-
'self, may I ever think of thee! May the daily
'Consciousness of my Weakness and Guilt lead
'my Thoughts to thee; and may every Thought
'of thee kindle in my Heart the most ardent Glow
'of Gratitude to thee, O thou Divine, Com-
'passionate Friend, Lover, and REDEEMER
'of Mankind!'

Whoever then he be that calls himself a Christian, that is, who professes to take the Gospel of Christ for a Divine Revelation, and the only Rule of his Faith and Practice; but at the same Time, pays a greater Regard to the Dictates of Men, than to the Doctrines of CHRIST; who loses Sight of that great *Example* of Christ, which should animate his Christian Walk, is unconcerned about his *Service, Honour* and *Interest*, and excludes the Consideration of his *Merits* and *Atonement*, from his Hope of Happiness; he forgets that he is a *Christian*;—— he does not consider in what Relation he stands to CHRIST, (which is one great Part of his Character,) and consequently discovers a great Degree of *Self-Ignorance*.

(3.) *Self-Knowledge* moreover implies a due Attention to the several *Relations* in which we stand to our *Fellow-Creatures:* and the Obligations that result from thence.

If we know ourselves, we shall remember the Con-

Condescension, Benignity, and Love that is due to Inferiors: the Affability, Friendship, and Kindness we ought to shew to Equals; the Regard, Deference, and Honour which belong to Superiors: and the Candour, Integrity, and Benevolence we owe to all.

The particular Duties requisite in these Relations are too numerous to be here mentioned. Let it suffice to say, that if a Man doth not well consider the several Relations of Life in which he stands to others, and does not take care to preserve the Decorum and Propriety of those Relations, he may justly be charged with *Self-Ignorance*.

And this is so evident in itself, and so generally allowed, that nothing is more common than to say, when a Person does not behave with due Decency towards his Superiors, such a one does not *understand himself*. But why may not this with equal Justice be said of those who act in an ill Manner towards their Inferiors? The Expression, I know, is not so often thus applied; but I see no Reason why it should not, since one is as common, and as plain an Instance of Self-Ignorance as the other. Nay, of the two, perhaps Men in general are more apt to be defective in their Duty and Behaviour towards those beneath them, than they are towards those that are above them. And the Reason seems to be, because an Apprehension of the Displeasure of their Superiors, and the detrimental Consequences which may accrue from thence, may be a Check upon them, and engage them to pay the juſt Regards which they expect. But there being no Check to reſtrain them from violating the Duties they owe to Inferiors, (from whoſe Diſpleaſure they have little to fear) they are more ready under certain Temptations to treat them in an unbecoming Manner. And as Wiſdom and Self-Knowledge will direct a Man to be particularly careful, leſt he neglect thoſe Duties he is moſt apt to forget; ſo as to the Duties he owes to Inferiors, in which he is moſt in danger of tranſgreſſing, he ought more ſtrongly to urge upon himſelf the indiſpenſible Obligations of Religion and Conſcience. And if he does not, but ſuffers himſelf through the Violence of ungoverned Paſſion to be tranſported into the Exceſſes of Rigour, Tyranny, and Oppreſſion, towards thoſe whom God and Nature have put into his Power, it is certain he does not *know himſelf*; is not acquainted with his own particular Weakneſs; is ignorant of the Duty of his Relation; and, whatever he may think of himſelf, hath not the true Spirit of Government; becauſe he wants the Art of Self-government. For he that is unable to govern himſelf, can never be fit to govern others.

Would we *know ourſelves* then, we muſt conſider ourſelves as *Creatures*, as *Chriſtians*, and as *Men*; and remember the Obligations, which as ſuch, we are under to GOD, to CHRIST, and our *Fellow-Men*; in the ſeveral *Relations* we bear to them, in order to maintain the Propriety, and fulfil the Duties of thoſe Relations.

CHAP.

CHAP. IV.

We must duly consider the Rank and Station of Life in which Providence has placed us, and what it is that becomes and adorns it.

III. A MAN *that knows himself, will deliberately consider and attend to the particular Rank and Station in Life in which Providence hath placed him; and what is the Duty and Decorum of that Station: what Part is given him to act, what Character to maintain; and with what Decency and Propriety he acts that Part, or maintains that Character.*

For a Man to assume a Character, or aim at a Part that does not belong to him, is Affectation. And whence is it that Affectation of any Kind appears so ridiculous, and exposes Men to universal and just Contempt; but because it is a certain Indication of Self-Ignorance? Whence is it that many seem so willing to be thought *Something, when they are Nothing*; and seek to excel in those Things in which they cannot; whilst they neglect those Things in which they may excel? Whence is it that they counter-act the Intention of Nature and Providence; that when these intended them one Thing, they would fain be another? Whence, I say, but from an Ignorance of themselves, the Rank of Life they are in, and of the Part and Character which properly belong to them?

It is a just Observation, and an excellent Document of a moral Heathen, that Human Life is a 'Drama, and Mankind the Actors, who
' have their several Parts assigned them by the
' Master of the Theatre, who stands behind
' the Scenes, and observes in what Manner every
' one acts. Some have a short Part allotted them,
' and some a long one: some a low, and some
' a high one. It is not he that acts the highest
' or most shining Part on the Stage, that comes
' off with the greatest Applause; but he that acts
' his Part best, whatever it be. To take Care
' then to act our respective Parts in Life well,
' is ours; but to choose what Part in Life we
' shall act, is not ours, but GOD's' (*r*).——
But a Man can never act his Part well, if he does not attend to it; does not know what becomes it; much less, if he affect to act another, which Nature never designed him. It is always *Self-Ignorance* that leads a Man to act out of Character.

Is it a mean and low Station of Life thou art in?——Know then, that Providence calls thee to the Exercise of Industry, Contentment, Submission,

---

(*r*) *Epictet. Enchir. cap.* 23.——Quomodo fabula, sic vita: non quàm diu, sed quàm bene acta sit, refert. *Sen. Ep.* 69. *ad fin. Life is a Stage-play; it matters not how long we act, so we act well.*——Non est bonum, vivere, sed benè vivere. *Id. de Benef. lib.* 3. *cap.* 31. *It is not Life, but living well,* that is the Blessing.

mission, Patience, Hope, and humble Dependence on Him, and a respectful Deference to thy Superiors. In this Way thou mayest shine through thine Obscurity; and render thyself amiable in the Sight of GOD and Man. And not only so, but find more Satisfaction, Safety, and Self-enjoyment, than they who move in a higher Sphere, from whence they are in Danger of falling.

But hath Providence called thee to act in a more publick Character, and for a more extensive Benefit to the World?——Thy first Care then ought to be, that thy Example, as far as its Influence reaches, may be an Encouragement to the Practice of universal Virtue. And next, to shine in those Virtues especially which best adorn thy Station; as, Benevolence, Charity, Wisdom, Moderation, Firmness, and inviolable Integrity: with an undismayed Fortitude to press through all Opposition in accomplishing those Ends which thou hast a Prospect and Probability of attaining for the apparent Good of Mankind.

And as Self-acquaintance will teach us what Part in Life we ought to act, so the Knowledge of that will shew us whom we ought to imitate, and wherein. We are not to take Example of Conduct from those who have a very different Part assigned them from ours; unless in those Things that are universally ornamental and exemplary. If we do, we shall but expose our Affectation and Weakness, and ourselves to Contempt for acting out of Character. For what is decent in in one may be ridiculous in another. Nor muſt we *blindly* follow thoſe who move in the ſame Sphere, and ſuſtain the ſame Character with ourſelves; but only in thoſe Things that are befitting that Character. For it is not the Perſon, but the Character, we are to regard; and to imitate him no farther than he keeps to that.

This Caution particularly concerns Youth, who are apt to imitate their Superiors very implicitly, and eſpecially ſuch as ſhine in the Profeſſion they themſelves are intended for; but, for Want of Judgment to diſtinguiſh what is fit and decent, are apt to imitate their very Foibles; which a Partiality for their Perſons makes them deem as Excellencies: and thereby they become doubly ridiculous, both by acting out of Character themſelves, and by a weak and ſervile Imitation of others in the very Things in which they do ſo too. To maintain a Character then with Decency, we muſt keep our Eye only upon that which is proper to it.

In fine, as no Man can excel in every Thing, we muſt conſider what Part is allotted us to act, in the Station in which Providence hath placed us, and to keep to that, be it what it will, and ſeek to excel in that only.

CHAP. V.

Every Man should be well acquainted with his own Talents and Capacities; and in what Manner they are to be exercised and improved to the greatest Advantage.

IV. A MAN cannot be said to know himself, till he is well acquainted with his proper Talents and Capacities; knows for what Ends he received them; and how they may be most fitly applied and improved for those Ends.

A wise and self-understanding Man, instead of aiming at Talents he hath not, will set about cultivating those he hath; as the Way in which Providence points out his proper Usefulness.

As in order to the Edification of the Church, the Spirit of God at first conferred upon the Ministers of it a great Variety of *Spiritual Gifts* *, so for the Good of the Community, GOD is pleased now to confer upon Men a great Variety of *Natural Talents*; and *every one hath his proper Gift of GOD; one after this Manner, another after that* †. And every one is to take Care *not to neglect,* but *to stir up the Gift of GOD* which *is in him* ‡. Because it was given him to be improved. And not only the Abuse, but the Neglect of it must hereafter be accounted for.

* 1 *Cor.* xii. 8—19.   † 1 *Cor.* vii. 7.   ‡ 1 *Tim.* iv. 14. 2 *Tim.* i. 6.

Witness the Doom of that *unprofitable Servant*, who *laid up his* single *Pound in a Napkin**; and of him who went and *hid his Talent in the Earth* †.

It is certainly a Sign of great *Self-ignorance*, for a Man to venture out of his Depth, or attempt any Thing he wants Opportunity or Capacity to accomplish. And therefore a wise Man will consider with himself, before he undertakes any Thing of Consequence, whether he hath Abilities to carry him through it, and whether the Issue of it is like to be for his Credit; lest he sink under the Weight he lays upon himself, and incur the just Censure of Rashness, Presumption, and Folly. See *Luke* xiv. 28,—32. (*s*).

It is no uncommon Thing for some who excel in one Thing, to imagine they may excel in every Thing. And not content with that Share

of

---

* *Luke* xix. 20, 24. † *Mat.* xxv. 25, 30.

(*s*) ——————————— ——————— Buccæ
Noscenda est mensura tuæ, spectandaque rebus
In summis, minimis,———————

Juv. Sat. 11.

———————verfate diu quid ferre recusant
Quid valeant Humeri.————————

Hor. de Art. Poet.

He that takes up a Burden that is too heavy for him, is in a fair Way to break his Back.

Ανθρωπε, πρωτον επισκεψαι, οποιον εςι το πραγμα· ειτα και την σαυτε φυσιν καταμαθε, ει δυνασαι βαςασαι. *Epict. Enchir. cap.* 36.

In every Business consider, first, what it is you are about; and then your own Ability, whether it be sufficient to carry you through it.

of Merit which every one allows them, are still catching at that which doth not belong to them. Why should a good Orator wish to be thought a Poet? Why must a celebrated Divine set up for a Politician? Or a Statesman affect the Philosopher? Or a Mechanick the Scholar? Or a wise Man labour to be thought a Wit? This is a Weakness that flows from *Self-ignorance*, and is incident to the greatest Men. Nature seldom forms a universal Genius; but deals out her Favours in the present State with a parsimonious Hand.——— Many a Man by this Foible hath weakened a well established Reputation (*t*).

(*t*)—————non omnia possumus omnes.   *Virg.*

Cæcilius, a famous Rhetorician of *Sicily*, who lived in the Time of *Augustus*, and writ a Treatise on the *Sublime*, (which is censured by *Longinus* in the Beginning of his) was a Man of a hasty and enterprising Spirit, and very apt to over-shoot himself on all Occasions; and particularly ventured out of his Depth in his *Comparison of Demosthenes and Cicero*: Whereupon Plutarch makes this sage and candid Remark. "If (saith he) it was a Thing obvious "and easy for every Man to know himself, possibly that Saying, "γνωθι σεαυτον, had not passed for a Divine Oracle." *Plut. Liv. Vol.* vii. *Pag.* 347.

CHAP. VI.

We must be well acquainted with our Inabilities, and those Things in which we are naturally deficient, as well as those in which we excel.

V. WE *must, in order to a thorough Self-acquaintance, not only consider our* Talents *and proper* Abilities, *but have an Eye to our* Frailties *and* Deficiencies, *that we may know where our* Weakness, *as well as our* Strength *lies.* —— Otherwise, like *Sampson*, we may run ourselves into infinite Temptations and Troubles.

Every Man hath a weak Side. Every wise Man knows where it is, and will be sure to keep a double Guard there.

There is some Wisdom in concealing a Weakness. This cannot be done, till it be first known, nor can it be known without a good Degree of Self-acquaintance.

It is strange to observe what Pains some Men are at to expose themselves; to signalize their own Folly; and to set out to the most publick View those Things which they ought to be ashamed to think should ever enter into their Character. But so it is; some Men seem to be ashamed of those Things which would be their Glory, whilst others *glory in their Shame* *.

The

---

* *Phil.* iii. 19.

The greatest Weakness in a Man is to publish his Weaknesses, and to appear fond to have them known. But Vanity will often prompt a Man to this; who, unacquainted with the Measure of his Capacities, attempts Things out of his Power, and beyond his Reach; whereby he makes the World acquainted with two Things to his Disadvantage, which they were ignorant of before; *viz.* his *Deficiency,* and his *Self-ignorance* in appearing so blind to it.

It is ill-judged (though very common) to be less ashamed of a Want of Temper than Understanding. For it is no real Dishonour or Fault in a Man to have but a small Ability of Mind, provided he have not the Vanity to set up for a Genius, (which would be as ridiculous, as for a Man of small Strength and Stature of Body, to set up for a Champion) because this is what he cannot help. But a Man may in a good measure correct the Fault of his natural Temper, if he be well acquainted with it, and duly watchful over it. And therefore to betray a prevailing Weakness of Temper, or an ungoverned Passion, diminishes a Man's Reputation much more than to discover a Weakness of Judgment or Understanding.———But what is most dishonourable of all is, for a Man at once to discover a great Genius and an ungoverned Mind. Because that Strength of Reason and Understanding he is Master of, gives him a great Advantage for the Government of his Passions. And therefore

fore his suffering himself notwithstanding to be governed by them, shews, that he hath too much neglected or misapplied his natural Talent; and willingly submitted to the Tyranny of those Lusts and Passions, over which Nature had furnished him with Abilities to have secured an easy Conquest.

A wise Man hath his Foibles as well as a Fool. But the Difference between them is, that the Foibles of the one are known to himself, and concealed from the World; the Foibles of the other are known to the World; and concealed from himself. The wise Man sees those Frailties in himself, which others cannot; but the Fool is blind to those Blemishes in his Character, which are conspicuous to every body else. Whence it appears, that *Self-Knowledge* is that which makes the main Difference between a wise Man and a Fool, in the Moral Sense of that Word.

CHAP. VII.

Concerning the Knowledge of our Constitutional Sins.

VI. SELF-ACQUAINTANCE *shews a Man the particular Sins he is most exposed and addicted to; and discovers not only what is* ridiculous, *but what is* criminal, *in his Conduct and Temper.*

A Man's

A Man's outward Actions are generally the plainest Index of his inward Dispositions. And by the allowed Sins of his Life, you may know the reigning Vices of his Mind. Is he addicted to Luxury and Debauch? Sensuality then appears to be his prevailing Taste. Is he given to Revenge and Cruelty? Choler and Malice then reign in his Heart. Is he confident, bold and enterprizing? Ambition appears to be the secret Spring. Is he sly and designing, given to Intrigue and Artifice? You may conclude, there is a natural Subtilty of Temper that prompts him to this; and this secret Disposition is criminal, in Proportion to the Degree in which these outward Actions, which spring from it, transgress the Bounds of Reason and Virtue.

Every Man hath something peculiar in the Turn or Cast of his Mind, which distinguishes him as much as the particular Constitution of his Body. And both these, *viz.* his particular Turn of Mind, and Constitution of Body, not only incline and dispose him to some Kind of Sins, more than to others, but render the Practice of certain Virtues much more easy*.

Now

---

* Men, with regard to their Bodies, and bodily Appetites, are pretty much alike; but with regard to their Souls, and their mental Tastes and Dispositions, they are often as different as if they were quite of another Species; governed by different Views, entertained with different Pleasures, animated with different Hopes, and affected by different Motives, and distinguished by as different Tempers and Inclinations, as if they were not of the same Kind. So that I am very ready to believe, that there is not a greater Diffe-

Now these Sins to which Men are commonly most inclined, and the Temptations to which they have least Power to resist, are, and not improperly, called their *Constitutional Sins*; their *peculiar* Frailties; and, in Scripture, their * *own Iniquities*, and the Sins which † *do most easily beset them* (*u*).

'As in the Humours of the Body, so in the
'Vices of the Mind, there is one predominant;
'which has an ascendant over us, and leads and
'governs us. It is in the Body of Sin, what the
'Heart is in the Body of our Nature; it begins
'to live first, and dies last. And whilst it lives,
'it communicates Life and Spirit to the whole
'Body of Sin; and when it dies, the Body of
'Sin

Difference between an Angel, and some of the best and wisest of Men; or between a Devil, and some of the worst and wickedest of Men, with regard to their Tempers and Dispositions, than there is between some Sort of Men, and some others. And what inclines me to this Sentiment is, considering the easy Transition which Nature always observes in passing from one Order or Kind of Beings to another, (which I have before taken notice of) together with the prodigious Difference there appears to be between some and others of the Human Species, almost in every Thing belonging to their Souls. For some there are, "in whom (as one expresses it) "one would think Nature had placed every Thing "the wrong Way;" depraved in their Opinions, unintelligible in their Reasoning, irregular in their Actions, and vitious in every Disposition. Whilst in some others we see almost every Thing amiable and excellent that can adorn and exalt the Human Mind, under the Disadvantages of Mortality.

* *Psalm* xviii. 23.   † *Heb.* xii. 1.

(*u*) η αμαρτια ευπερισατ⊙, *the well circumstanced Sin.*

'Sin expires with it. It is the Sin to which our
'Constitution leads, our Circumstances betray,
'and Custom inslaves us; the Sin to which not
'our Virtues only, but Vices too, lower their
'Topsail, and submit; the Sin, which when we
'would impose upon GOD and our Consciences
'we excuse and disguise with all imaginable Ar-
'tifice and Sophistry; but, when we are sincere
'with both, we oppose first, and conquer last.
'It is, in a Word, the Sin which reigns and
'rules in the Unregenerate, and too often alarms
'and disturbs (ah! that I could say no more) the
'Regenerate' (*w*).

Some are more inclined to the Sins of the *Flesh*; Sensuality, Intemperance, Uncleanness, Sloth, Self-indulgence, and Excess in animal Gratifications. Others to the Sins of the *Spirit*; Pride, Malice, Coveteousness, Ambition, Wrath, Revenge, Envy, &c. And I am persuaded there are few, but, upon a thorough Search into themselves, may find that some one of these Sins hath ordinarily a greater Power over them than the rest. Others often observe it in them, if they themselves do not. And for a Man not to know his predominant Iniquity is great Self-ignorance indeed; and a Sign that he has all his Life lived far from Home; because he is not acquainted with *that* relating to himself, which every one, who is but half an Hour in his Company, perhaps may be able to inform him of. Hence proceeds

(*w*) See Dr. *Lucas*'s Sermons, *Vol.* i. *pag.* 151.

proceeds that extreme Weakness which some discover in censuring others, for the very same Faults they are guilty of themselves, and perhaps in a much higher Degree; on which the Apostle *Paul* animadverts, *Rom.* ii. 1. (*x*).

It must be owned, it is an irksome and disagreeable Business for a Man to turn his own Accuser; to search after his own Faults, and keep his Eye upon that which gives him Shame and Pain to see. It is like tearing open an old Wound. But it is better to do this, than to let it mortify. The Wounds of the Conscience, like those of the Body, cannot be well cured till they are searched to the Bottom; and they cannot be searched without Pain. A Man, that is engaged in the Study of himself, must be content to know the worst of himself (*y*).

Do not therefore shut your Eyes against your darling Sin, or be averse to find it out. Why should you study to conceal or excuse it; and fondly
cherish

(*x*) Quis tulerit *Gracchos* de seditione querentes? *Clodius* accusat Mæchos? *Catalina Cathegum*?

Juv. Sat. 12.

(*y*) O nimis gravis angustia! Si me inspicio, non tolero meipsum: si non inspicio, nescio meipsum. Si me considero, terret me Facies mea: si me non considero, fallit me Damnatio mea. Si me video, Horror est intolerabilis: si non video, Mors est inevitabilis. ——— *O grievous Streight! If I look into myself, I cannot endure myself: If I look not into myself, I cannot know myself. If I consider myself, my own Face affrights me: If I consider not myself, my Damnation deceives me. If I see myself, my Horror is intolerable: If I see not myself, Death is unavoidable.* Anselme.

cherish that Viper in your Bosom?———'Some
'Men deal by their Sins, as some Ladies do by
'their Persons. When their Beauty is decayed,
'they seek to hide it from themselves by false
'Glasses, and from others by Paint. So, many
'seek to hide their Sins from themselves by false
'Glosses, and from others by Excuses, or false
'Colours' (z). But the greatest Cheat they put
upon themselves. * *They that cover their Sins
shall not prosper.* It is dangerous Self-flattery
to give soft and smoothing Names to Sins, in
order to disguise their Nature. Rather lay your
Hand upon your Heart, and † *thrust it into your
Bosom,* though it *come out* (as *Moses*'s did) *leprous
as Snow* (a).

And to find out our most beloved Sin, let us
consider what are those worldly Objects or Amusements which give us the highest Delight; this, it
is

(z) *Baxter.*

* *Prov.* xxviii. 13. † *Exod.* iv. 6.

(a) Initium est salutis, notitia peccati: nam qui peccare se nescit, corrigi non vult. Deprehendas te oportet, antequàm emendes. Quidam vitiis gloriantur. Tu existimas aliquid de Remedio cogitare, qui mala sua virtutum loco numerant? Ideo quantum potes teipsum coargue: Inquire in te: Accusatoris primum partibus fungere, deinde Judicis, novissimè Deprecatoris. Aliquando te offende. *Sen. Epist.* 28. ——— *The Knowledge of Sin is the first Step towards Amendment: for he that does not know he hath offended, is not willing to be reproved. You must therefore find out yourself, before you can amend yourself. Some glory in their Vices. And do you imagine they have any Thought about reforming, who place their very Vices in the room of Virtues? Therefore reprove thyself: search thyself very narrowly. First turn Accuser to thyself, then a Judge, and then a Suppliant. And dare for once to displease thyself.*

is probable, will lead us directly to some one of our darling Iniquities, if it be a Sin of Commission: and what are those Duties which we read or hear of from the Word of GOD, to which we find ourselves most disinclined? And this, in all likelihood, will help us to detect some of our peculiar Sins of Omission; which without such previous Examination we may not be sensible of. And thus we may make a Proficiency in one considerable Branch of *Self-Knowledge* (*b*).

CHAP.

(*b*) Et hoc ipsum Argumentum est in melius translati animi, quod Vitia sua, quæ adhuc ignorabat, videt. *Sen. Epist.* 6. —— *It is a good Argument of a reformed Mind, that it sees those Vices in itself, which it was before ignorant of.*

A Man's predominant Sin usually arises out of his predominant Passion; which therefore he should diligently observe. The Nature and Force of which is beautifully described by a late great Master of English Verse.

On different Senses different Objects strike,
Hence different Passions more or less enflame,
As strong or weak, the Organs of the Frame;
And hence one Master-Passion in the Breast,
Like *Aaron*'s Serpent swallows up the rest.
Nature its Mother, Habit is its Nurse;
Wit, Spirit, Faculties, but make it worse;
Reason itself but gives it Edge and Power,
As Heaven's blest Beam turns Vinegar more sowr.
Ah! If she lend not Arms as well as Rules,
What can she more than tell us we are Fools?
Teach us to mourn our Nature, not to mend,
A sharp Accuser, but a helpless Friend!

Pope's Essay on Man.

CHAP. VIII.

The Knowledge of our most dangerous Temptations, necessary to Self-Knowledge.

VII. A MAN that rightly knows himself, is acquainted with his peculiar Temptations; *and knows when, and in what Circumstances, he is in the greatest Danger of transgressing.*

Reader, if ever you would *know yourself*, you must examine this Point thoroughly. And if you have never yet done it, make a Pause when you have read this Chapter, and do it now. Consider in what Company you are most apt to lose the Possession and Government of yourself; on what Occasions you are apt to be most vain and unguarded, most warm and precipitant. Flee that Company, avoid those Occasions, if you would keep your Conscience clear. What is it that robs you most of your Time and your Temper? If you have a due Regard to the Improvement of the one, and the Preservation of the other, you will regret such a Loss; and shun the Occasions of it, as carefully as you would a Road beset with Robbers.

But especially must you attend to the Occasions which most usually betray you into your favourite Vices; and consider the Spring from whence they arise,

arise, and the Circumstances which most favour them. They arise doubtless from your natural Temper, which strongly disposes and inclines you to them. *That* Temper then, or particular Turn of Desire, must be carefully watched over as a most dangerous Quarter. And the Opportunities and Circumstances which favour those Inclinations must be resolutely avoided, as the strongest Temptations. For the Way to subdue a criminal Inclination is, first, to avoid the known Occasions that excite it; and then, to curb the first Motions of it (*c*). And thus having no Opportunity of being indulged, it will of itself in Time lose its Force, and fail of its wonted Victory.

The surest Way to conquer, is sometimes to decline a Battle; to weary out the Enemy, by keeping him at Bay. *Fabius Maximus* did not use this Stratagem more successfully against *Hannibal* than a Christian may against his peculiar Vice, if he be but watchful of his Advantages. It is dangerous to provoke an unequal Enemy to the Fight, or to run into such a Situation, where we cannot expect to escape without a disadvantageous Encounter.

It is of unspeakable Importance, in order to *Self-Knowledge* and *Self-Government*, to be acquainted with all the Accesses and Avenues to Sin, and to observe which Way it is that we ourselves too often approach it; and to set Reason and Conscience to guard

---

(*c*) Principiis obsta: sero medicina paratur
  Cum mala per longas invaluere moras.   *Ovid.*

guard thofe Paffes, thofe ufual Inlets to Vice, which if a Man once enters, he will find a Retreat extremely difficult (*d*).

'Watchfulnefs, which is always neceffary, is
'chiefly fo when the firft Affaults are made. For
'then the Enemy is moft eafily repulfed; if we
'never fuffer him to get within us, but upon the
'very firft Approach draw up our Forces, and
'fight him without the Gate. And this will be
'more manifeft, if we obferve by what Methods
'and Degrees Temptations grow upon us.———
'The firft Thing that prefents itfelf to the Mind
'is a plain fingle *Thought*; this ftreight is im-
'proved into a ftrong *Imagination*; that again en-
'forced by a fenfible *Delight*; then follow evil
'*Motions*; and when thefe are once ftirred there
'wants nothing but the *Affent* of the *Will*, and
'then the Work is finifhed. Now the firft Steps
'to this are feldom thought worth our Care;
'fometimes not taken Notice of; fo that the Ene-
'my is frequently got clofe up to us, and
'even within our Trenches, before we obferve
'him (*e*).'

As Men have their particular Sins, which do *moft eafily befet them*; fo they have their particu-

---

(*d*) ——————— me veftigia terrent
   Omnia te adverfum fpectantia, nulla retrorfum.

Hor.

——————— Facilis defcenfus averni.
Sed revocare gradum, &*c*.

Virg.

(*e*) *Stanhope's Thomas à Kempis, pag. 22.*

lar Temptations, which do most *easily overcome* them. *That* may be a very great Temptation to one, which is none at all to another. And if a Man does not know what are his greatest Temptations, he must have been a great Stranger indeed to the Business of Self-Employment.

As the subtle Enemy of Mankind takes Care to draw Men gradually into Sin, so he usually draws them by degrees into Temptation. As he disguises the Sin, so he conceals the Temptation to it; well knowing, that were they but once sensible of their Danger of Sin, they would be ready to be on their Guard against it. Would we know ourselves thoroughly then, we must get acquainted not only with our most usual Temptations, that we be not unawares drawn into *Sin*, but with the previous Steps and preparatory Circumstances, which make Way for those Temptations, that we be not drawn unawares into the *Occasions* of Sin; for those Things which lead us into Temptations are to be considered as Temptations, as well as those which immediately lead us into Sin. And a Man that knows himself will be aware of his remote Temptations, as well as the more immediate ones; *e. g.* If he find the Company of a passionate Man is a Temptation (as *Solomon* tells us it is, *Prov.* xxii. 24, 25.) he will not only avoid it, but those Occasions that may lead him into it. And the Petition in the *Lord's-Prayer* makes it as much a Man's Duty to be upon his Guard *against* Temptation, as under it.

it. Nor can a Man pray from his Heart that GOD would not *lead him into Temptation*, if he take no Care himself to avoid it.

CHAP. IX.

Self-Knowledge discovers the secret Prejudices of the Heart.

VIII. ANOTHER *important Branch of Self-Knowledge is, for a Man to be acquainted with his* own Prejudices; or those secret Prepossessions of his Heart, which, though so *deep* and *latent*, that he may not be sensible of them, are often so *strong* and *prevalent*, as to give a mighty, but *imperceptible* Bias to the Mind.

There is no one Particular that I know of wherein Self-Knowledge more eminently consists than it does in this. It being therefore so essential a Branch of my Subject, and a Point to which Men seldom pay an Attention equal to its Importance, I beg leave to treat it with a little more Precision.

These Prejudices of the Human Mind may be considered with regard to *Opinions, Persons,* and *Things.*

(1.) With regard to *Opinions.*

It is a common Observation, but well expressed by a late celebrated Writer, ' that we set out ' in Life with such poor Beginnings of Know- ' ledge, and grow up under such Remains of Su- ' perstition and Ignorance, such Influences of
' Company

'Company and Fashion, such Insinuations of
'Pleasure, &c. that it is no Wonder, if Men
'get Habits of thinking only in one Way; that
'these Habits in Time grow rigid and confirmed;
'and so their Minds come to be overcast with
'thick Prejudices, scarce penetrable by any Ray
'of Truth, or Light of Reason *(f)*.'

There is no Man but is more attached to one particular Set or Scheme of Opinions in Philosophy, Politics and Religion, than he is to another; I mean if he hath employed his Thoughts at all about them. The Question we should examine then is; how came we by these Attachments? Whence are we so fond of these particular Notions? Did we come fairly by them? or were they imposed upon us, and dictated to our easy Belief, before we were able to judge of them? This is most likely. For the Impressions we early receive generally grow up with us, and are those we least care to part with. However, which Way soever we came by them, they must be re-examined, and brought to the *Touchstone* of sound Sense, solid Reason, and plain Scripture. If they will not bear this after hard rubbing, they must be dismissed, as no genuine Principles of Truth, but as Counterfeits imposed upon us under the Guise and Semblance of it.

And as *Reason* and *Scripture* must discover our Prejudices to us, so they only can help us to get rid of them. By these are we to rectify, and to these are we to conform, all our Opinions and
Senti-

(f) See *Religion of Nature delin.* Pag. 129.

Sentiments in Religion, as our only Standard, exclusive of all other Rules, Light, or Authority, whatsoever.

And Care must further be taken that we do not, make Scripture and Reason bend and buckle to our Notions; which will rather confirm our Prejudices than cure them. For whatever cannot evidently be proved, without the Help of overstrained Metaphors, and the Arts of Sophistry, is much to be suspected; which used to make Archbishop *Tillotson* say, *Non amo argutias in Theologiâ*; I do not love Subtilties in Divinity. But,

(2.) The Human Mind is very apt to be prejudiced either for or against certain *Persons*, as well as certain Sentiments. And as Prejudice will lead a Man to *talk* very unreasonably with regard to the latter, so will it lead him to *act* as unreasonably with regard to the former.

What is the Reason, for Instance, that we cannot help having a more hearty Affection for some Persons than others? Is it from a Similarity of Taste and Temper? Or something in their Address, that flatters our Vanity? Or something in their Humour that hits our Fancy? Or something in their Conversation, that improves our Understanding? Or a certain Sweetness of Disposition, and Agreeableness of Manner, that is naturally engaging? Or from Benefits received or expected from them? Or from some eminent and distinguished Excellency in them? Or from none of these; but something else, we cannot tell what?

what?———Such Sort of Enquiries will shew us whether our Esteem and Affections be rightly placed: or flow from mere Instinct, blind Prejudice, or something worse.

And so on the other Hand, with regard to our Disaffection towards any one, or the Disgust we have taken against him; if we would know ourselves, we must examine into the Bottom of this; and see not only what is the pretended, but true Cause of it: whether it be justifiable, and our Resentments duly proportioned to it.———Is his Manner of Thinking, Talking, and Acting, quite different from mine, and therefore what I cannot approve? Or have I received some real Affront or Injury from him? Be it so, my continued Resentment against him, on either of these Accounts, may be owing, notwithstanding, more to some unreasonable Prejudice in *me*, than to any real Fault in *him*.

For as to the former: His Way of *Thinking*, *Talking*, and *Acting*, may possibly be juster than my own; which the mere Force of Custom and Habit only makes me prefer to his. However, be it ever so wrong, he may not have had the same Advantage of improving his Understanding, Address, and Conduct, as I have had; and therefore his Defects herein are more excusable. And he may have many other Kind of Excellencies which I have not.———' But he is not only ig-
' norant and unmanner'd, but unsufferably vain,
' conceited and overbearing at the same Time.'
———Why,

—— Why, *that* perhaps he cannot help. It is the Fault of his Nature. He is the Object of Pity rather than Resentment. And had I such a Disposition by Nature, I should perhaps, with all my Self-improvement, find it a difficult Thing to manage. And therefore, tho' I can never chuse such a one for an agreeable Companion, yet I ought not to harbour a Dislike to him, but love, and pity, and pray for him, as a Person under a great Misfortune; and be thankful that I am not under the same.——' But he is quite blind to this ' Fault of his Temper, and does not appear to be ' in the least sensible of it.' —— Why, that is a greater Misfortune still; and he ought to be the *more* pitied.

And as to the other pretended Ground of Disgust, ' he hath often *offended* and *injured* me.' Let me consider, (1.) *Whether any Offence was really intended*; whether I do not impute that to ill Nature, which was only owing to ill Manners; or that to Design, which proceeded only from Ignorance. Do I not take Offence before it is given? If so, the Fault is mine, and not his. And the Resentment I have conceived against him, I ought to turn upon myself (*g*).——Again, (2.) *Did I not provoke him to it, when I knew his Temper?* The Fault is still my own. I did, or

D 4 might

(*g*) For every Trifle scorn to take Offence;
    That always shews great Pride or little Sense.
    Good Nature and good Sense must always join;
    To err is Human, to forgive Divine.

Pope.

might know the Pride, Paſſion, or Perverſeneſs of his Nature; why then did I exaſperate him? A Man that would needleſsly rouſe a Lion, muſt not expect always to come off ſo favourably as the Hero of *La Mancha*.———But, (3.) Suppoſe I were not the Aggreſſor; yet, *how came I into his Company?* Who led me into the Temptation? He hath acted according to his Nature in what he hath done; but I have not acted according to my Reaſon, in laying myſelf ſo open to him. I knew him; why did I not ſhun him, as I would any other dangerous Animal that does Miſchief by Inſtinct? If I muſt needs put my Finger into a Waſp's Neſt, why ſhould I blame them for ſtinging me?———Or, (4.) If I could not avoid his Company, *why did I not arm myſelf?* Why did I venture defenceleſs into ſo much Danger? Or, (5.) Suppoſe he hath done me a real and undeſerved Injury, without my Fault or Provocation; yet *does not my Diſcontent aggravate it?* Does it not appear greater to me, than it does to any body elſe? or than it will to *me*, after the preſent Ferment is over?——— And (Laſtly,) after all, *muſt I never forgive?* How ſhall I be able to repeat the Lord's Prayer, or read our Saviour's Comment upon it, *Mat.* vi. 14, 15. with an unforgiving Temper? Do I not hope to be forgiven *Ten thouſand Talents*; and cannot I forgive my *Fellow-Servant Thirty-Pence?* When I know not but he hath repented, and GOD hath forgiven him, whoſe Forgiveneſs I want

want infinitely more than my greatest Enemy does mine *.

Such Considerations are of great Use to soften our Prejudices against Persons; and at once to discover the true Spring, and prevent the bad Effects of them. And happy would it be for a Christian, could he but call to Mind and apply to his Relief, half the good Things which that excellent Heathen Emperor and Philosopher *Marcus Antoninus* could say upon this Subject. Some of which I have, for the Benefit of the English

---

* A Man despises me: what then? Did he know me more, he would perhaps despise me more. But I know myself better than he can know me; and therefore despise myself more. And though his Contempt in this Instant may be groundless, yet in others it would be but too well founded. I will therefore not only bear with, but forgive it.—*Contemnendus est ipse contemptus,* saith *Seneca.* But such *retorted Scorn* is more becoming the Character of a *Stoic* than a *Christian.*

It has been reckoned a wise and witty Answer which one of the Philosophers returned to his Friend, who advised him to revenge an Injury that had been done him: " What (says he) if an Ass " kicks me, must I needs kick him again? " And perhaps there is more *Wit* than *Wisdom* in that Reply. It seems indeed to carry in it something of a true Greatness of Mind; but does it not at the same Time discover a kind of haughty and contemptuous Spirit? The Truth is, (as a Judicious Writer observes upon it) " it " is at best but a lame and mishapen Charity; it has more of " Pride than Goodness. We should learn of the holy *Jesus,* who " was not only *meek,* but *lowly.* We should contemn the *Injury,* " and pity the *Weakness*; but should not disdain or despise the " *Persons* of our Enemies. *Charity vaunteth not herself, is not puff-* " *ed up, doth not behave itself unseemly.*" See *Scougal's Duty of loving our Enemies.*

Reader extracted, and thrown into the Margin (*h*).

(3.) The

(*h*) In the Morning remember to say to thyself; this Day perhaps I may meet with some impertinent, ungrateful, peevish, tricking, envious, churlish Fellow. Now all these ill Qualities in them proceed from their Ignorance of Good and Evil. And since I am so happy as to understand the natural Beauty of a good Action, and the Deformity of an ill One; and since the Person that disobliges me is of near Kin to me; and tho' not just of the same Blood and Family, yet of the same divine Extract as to his Mind; and finally, since I am convinced that no one can do me a real Injury, because he cannot force me to do a dishonest Thing; for these Reasons I cannot find in my Heart to hate him, or so much as to be angry with him. *Marc. Anton. Medit.* Book 2. § 1.

You are just taking Leave of the World; and have you not yet learned to be Friends with every body? And that to be an honest Man, is the only Way to be a wise one? *Id.* Book 4. § 37.

To expect an Impossibility is Madness; now it is impossible for ill Men not to do ill Things. *Id.* Book 5. § 17.

It is the Privilege of Human Nature above Brutes to love those that offend us; in order to this consider, (1.) That the offending Party is of Kin to you; (2.) That he acts thus, because he knows no better; (3.) He may have no Design to offend you; (4.) You will both of you quickly be in your Graves; but above all, (5.) You have received no Harm from him. For your Mind or Reason is the same it was before. *Id.* Book 7. § 22.

Think upon your last Hour, and do not trouble yourself about other People's Faults, but leave them there where they must be answered for. *Id.* Book 7. § 29.

Do not return the Temper of ill-natured People upon themselves, nor treat them as they do the rest of Mankind. *Id.* Book 7. § 55.

Tho' the Gods are immortal, yet they not only patiently bear with a wicked World through so many Ages; but what is more, liberally provide for it: and are you, who are just going off the Stage, weary with bearing, tho' you are one of those unhappy Mortals yourself? *Id.* Book 7. § 70.

Never disturb yourself; for Men will do the same untoward Actions over again, tho' you burst with Spleen. *Id.* Book 8. § 4.

Reform

(3.) The Mind is apt to be prejudiced against or in favour of certain *Things* and *Actions*, as well as certain Sentiments and Persons.

Do you not sometimes find dull disagreeable Ideas annexed to certain Places, Seasons, or Employments, which give you a secret Aversion to them? These arise from the Remembrance of some unpleasing Incidents you have heretofore met with, and which you apprehend may again befal you on such Occasions. But they are often nothing more than the mere Misrepresentations of Fancy; and ought to be repelled, because they will be apt to lead you to neglect the Duties of your Character.

If therefore you find in yourself a secret Disinclination to any particular Action or Duty, and the Mind begins to cast about for Excuses and Reasons to justify the Neglect of it, consider the Matter well: Go to the Bottom of that Reluctance;

Reform an injurious Person if you can; if not, remember your Patience was given you to bear with him. That the Gods patiently bear with such Men, and sometimes bestow upon them Health, and Fame, and Fortune. *Id* Book 9. § 11.

When People treat you ill, and show their Spite, and slander you, enter into their little Souls, go to the Bottom of them, search their Understandings; and you will soon see, that nothing they may think or say of you need give you one troublesome Thought. *Id.* Book 9. § 27.

That is the best Thing for a Man which GOD sends him; and that is the best Time when he sends it. *Id.* Book 10. § 2.

It is sometimes a hard Matter to be certain, whether you have received ill Usage or not; for Mens Actions oftentimes look worse than they are: and one must be thoroughly informed of a great many Things, before he can rightly judge *Id.* Book 11. § 18.

tance; and search out what it is that gives the Mind this Aversion to it. Whether it be the Thing or Action itself, or some discouraging Circumstances that may attend it; or some disagreeable Consequences that may possibly flow from it; or your supposed unfitness for it at present. Why, all these Things may be only imaginary. And to neglect a plain and positive Duty upon such Considerations, shows that you are governed by Appearances more than Realities, by Fancy more than Reason, and by Inclination more than Conscience.

But let Fancy muster up all the discouraging Circumstances, and set them in the most formidable Light, to bar your Way to a supposed Duty; for instance, 'It is very difficult, I want Ca-
'pacity, at least am so indisposed to it at present,
' that I shall make nothing of it; and then it
' will be attended with Danger to my Person, Re-
' putation

Consider how much more you often suffer from your Anger and Grief, than from those very Things for which you are angry and grieved. *Id. Book* 11. § 18.

When you fancy any one hath transgressed, say thus to yourself: 'How do I know it is a Fault? But admit it is, it may be his Con-
' science hath corrected him: and then he hath received his Pu-
' nishment from himself.' *Id. Book* 12. § 16.

To these I shall add two more Quotations out of the Sacred Writings, of incomparable greater Weight and Dignity than any of the forementioned. *Prov.* xix. 11. *The Discretion of a Man deferreth his Anger: and it is his Glory to pass over a Transgression.* Rom. xii. 20. 21. *If thine Enemy hunger, feed him; if he thirst, give him Drink: for in so doing thou shalt heap Coals of Fire on his Head. Be not overcome of Evil, but overcome Evil with Good.*

'putation or Peace; and the Opposition I am like
'to meet with is great, &c.' But after all, is the
Call of Providence clear? Is the Thing a plain
Duty? such as Reason, Conscience, and Scripture;
your Office, Character, or Personal Engagements
call upon you to discharge? If so, all the aforesaid
Objections are vain and delusive; and you have
nothing to do, but to summon your Courage, and
in Dependance on Divine Help, to set about the
Business immediately and in good Earnest, and in
the best and wisest Manner you can; and you
may depend upon it, you will find the greatest
Difficulty to lie only in the first Attempt; these
frightful Appearances to be all visionary, the mere
Figments of Fancy, turning Lambs into Lions,
and Mole-hills into Mountains; and that nothing
but Sloth, Folly and Self-indulgence, thus set your
Imagination on work to deter you from a plain
Duty. Your Heart would deceive you, but you
have found out the Cheat, and do not be imposed upon (*i*).

Again, suppose the Thing done; consider how
it will look then. Take a View of it as past; and
whatever Pains it may cost you, think whether it
will not be abundantly recompensed by the inward
Peace and Pleasure, which arise from a Consciousness of having acted right. It certainly will.
And

(*i*) "The wise and prudent conquer Difficulties,
"By daring to attempt them. Sloth and Folly
"Shiver and shrink at Sight of Toil and Danger,
"And make th' Impossibility they fear." *Rowe.*

And the Difficulties you now dread will enhance your future Satisfaction (*k*). But think again how you will bear the Reflections of your own Mind if you wilfully neglect a plain and necessary Duty; whether this will not occasion you much more Trouble than all the Pains you might be at in performing it. And a wise Man will always determine himself by the End; or by such a retrospective View of Things, considered as past.

Again, on the other Hand, if you find a strong *Propension* to any particular Action, examine *that* with the like Impartiality. Perhaps it is what neither your Reason nor Conscience can fully approve. And yet every Motive to it is strongly urged, and every Objection against it slighted. Sense and Appetite grow importunate and clamorous, and want to lead, while Reason remonstrates in vain. But turn not aside from that faithful and friendly Monitor, whilst with a low, still Voice, she addresses you in this soft but earnest Language.———' Hear me, I beseech you,
' but this one Word more. The Action is *in-*
' *deed* out of Character; what I shall never ap-
' prove. The Pleasure of it is a great deal over-
' rated; you will certainly be disappointed. It is
' a false Appearance that now deceives you. And
' what will you think of yourself when it is past,
' and you come to reflect seriously on the Mat-
' ter? Believe it, you will then wish you had
' taken me for your Counsellor, instead of those
' Ene-

(*k*) ———forsan et hæc olim meminisse juvabit.

Virg.

'Enemies of mine, your Lusts and Passions, which have so often misled you, tho' you know I never did.'——

Such short Recollections as these, and a little Leisure to take a View of the Nature and Consequences of Things or Actions, before we reject or approve them, will prevent much false Judgment and bad Conduct; and by Degrees wear off the Prejudices which Fancy has fixed in the Mind either for or against any particular Action; teach us to distinguish between Things and their Appearances; strip them of those false Colours that so often deceive us; correct the Sallies of the Imagination, and leave the Reins in the Hand of Reason.

Before I dismiss this Head, I must observe, that some of our strongest Prejudices arise from an excessive *Self-esteem*, or too great a Complacency in our own good Sense and Understanding. *Philautus* in every Thing shews himself well satisfied with his own Wisdom: which makes him very impatient of Contradiction, and gives him a Distaste to all who shall presume to oppose their Judgment to his in any Thing. He had rather persevere in a Mistake than retract it, lest his Judgment should suffer; not considering that his Ingenuity and good Sense suffer much more by such Obstinacy. The Fulness of his Self-sufficiency makes him blind to those Imperfections which every one can see in him but himself. So that, however wise, sincere and friendly, however gentle and seasonable your Remonstrance may be, he takes it immediately to proceed from Ill-nature or Ignorance in *you*, but from no Fault in *him*.

Seneca, I remember, tells us a remarkable Story, which very well illustrates this Matter.—— Writing to his Friend *Lucilius*, 'My Wife (says
' he) keeps *Harpastes* in her House still, who, you
' know, is a Sort of Family-Fool, and no small
' Incumbrance upon us. For my Part, I am far
' from taking any Pleasure in such Prodigies. If
' I have a Mind to divert myself with a Fool, I
' have not far to go for one; I can laugh at my-
' self. This silly Girl, all on a sudden, lost her
' Eye-sight; and (which perhaps may seem incre-
' dible, but it is very true) she does not know she
' is blind; but is every now and then desiring her
' Governess to lead her abroad, saying the House
' is dark. ——— Now what we laugh at in this
' poor Creature, you may observe happens to us all.
' No Man knows that he is covetous, or insatiable.
' Yet with this Difference; the Blind seek some-
' body to lead them, but we are content to wan-
' der without a Guide. —— But why do we thus
' deceive ourselves? The Disease is not without
' us, but fixed deep within. And therefore is the
' Cure so difficult, because we do not know that
' we are sick (*l*).'

(*l*) *Sen. Epist.* 51.

CHAP.

CHAP. X.

The Necessity and Means of knowing our Natural Temper.

IX. ANOTHER *very important Branch of Self-Knowledge is, the Knowledge of those Governing Passions or Dispositions of the Mind, which generally form what we call a Man's* Natural Temper.

The Difference of Natural *Tempers* seems to be chiefly owing to the different Degrees of Influence the several Passions have upon the Mind. *e. g.* If the Passions are eager and soon raised, we say the Man is of a *warm* Temper; if more sluggish and slowly raised, he is of a *cool* Temper; according as Anger, Malice or Ambition prevail, he is of a *fierce, churlish,* or *haughty* Temper; the Influence of the softer Passions of Love, Pity and Benevolence, forms a *sweet, sympathising* and *courteous* Temper; and when all the Passions are duly poised, and the milder and pleasing ones prevail, they make what is commonly called a quite *good-natured* Man.

So that it is the Prevalence or Predominance of any particular Passion which gives the Turn or Tincture to a Man's Temper, by which he is distinguished, and for which he is loved or esteemed, or shunned and despised by others.

Now what this is, those we converse with are soon sensible of. They presently see the Fault of our

our Temper, and order their Behaviour accordingly. If they are wife and well mannered, they will avoid striking the String which they know will jarr and raise a Discord within us. If they are our Enemies, they will do it on Purpose to set us on tormenting ourselves. And our Friends we must suffer sometimes with a gentle Hand to touch it, either by Way of pleasant Raillery or faithful Advice.

But a Man must be greatly unacquainted with himself, if he is ignorant of his predominant Passion, or distinguishing Temper, when every one else observes it. And yet how common is this Piece of Self-ignorance? The two Apostles *Peter* and *John* discovered it in that very Action, wherein they meant to express nothing but a hearty Zeal for their Master's Honour; which made him tell them, *that they knew not what Manner of Spirit they were of,* Luke ix. 5. *i. e.* that, instead of a Principle of Love and genuine Zeal for him, they were at that Time governed by a Spirit of Pride, Revenge and Cruelty; and yet knew it not. And that the Apostle *John* should be liable to this Censure, whose Temper seemed to be all Love and Sweetness, is a memorable Instance how difficult a Thing it is for a Man at all Times to know his own Spirit; and that *that* Passion, which seems to have the least Power over his Mind, may on some Occasions insensibly gain a criminal Ascendant there.

The

The Necessity of a perfect Knowledge of our reigning Passions appears further from hence; that they not only give a Tincture to the Temper, but to the Understanding also; and throw a strong Bias on the Judgment. They have much the same Effect upon the Eye of the Mind, as some Distempers have upon that of the Body. If they do not put it out they weaken it; or throw false Colours before it, and make it form a wrong Judgment of Things. And, in short, are the Source of those forementioned Prejudices, which so often abuse the Human Understanding.

Whatever the different Passions themselves that reign in the Mind may be owing to, whether to the different Texture of the bodily Organs, or the different Quantity or Motion of the Animal Spirits, or to the native Turn and Cast of the Soul itself; yet certain it is, that Men's different Ways of thinking are much according to the Predominance of their different Passions; and especially with regard to Religion. Thus, *e. g.* we see melancholy People are apt to throw too much Gloom upon their Religion, and represent it in a very uninviting and unlovely View, as all Austerity and Mortification; whilst they, who are governed by the more gay and chearful Passions, are apt to run into the other Extreme, and too much to mingle the Pleasures of Sense with those of Religion; and are as much too lax, as the other too severe. And thus by the Prejudice or Bias of their respective Passions, or the Force of their

Natural

Natural Temper, they are led into different Mistakes.

'So that would a Man know himself, he must
'study his Natural Temper; his constitutional
'Inclinations, and favourite Passions; for by these
'a Man's Judgment is easily perverted, and a
'wrong Bias hung upon his Mind: These are the
'Inlets of Prejudice; the unguarded Avenues of
'the Mind, by which a thousand Errors and se-
'cret Faults find Admission, without being ob-
'served or taken Notice of (*m*).

And that we may more easily come at the Knowledge of our predominant Affections, let us consider what outward Events do most impress and move us; and in what Manner. What is it that usually creates the greatest Pain or Pleasure in the Mind?——As for *Pain*; a *Stoic* indeed may tell us, 'that we must keep Things at a Distance; let nothing that is outward come within us; let Externals be Externals still.' But the human Make will scarce bear the Rigour of that Philosophy. Outward Things, after all, will impress and affect us. And there is no Harm in this, provided they do not get the Possession of us, overset our Reason, or lead us to act unbecoming a Man or a Christian. And one Advantage we may reap from hence is, the Manner or Degree in which outward Things impress us, may lead us into a better Acquaintance with *ourselves*, discover

---

(*m*) *Spectat.* Vol. vi. No. 899.

discover to us our weak Side, and the Passions which most predominate in us.

Our *Pleasures* will likewise discover our reigning Passions, and the true Temper and Disposition of the Soul. If it be captivated by the Pleasures of Sin, it is a Sign its prevailing Taste is very vicious and corrupt; if with the Pleasures of Sense, very low and sordid; if imaginary Pleasures, and the painted Scenes of Fancy and Romance do most entertain it, the Soul hath then a trifling Turn; if the Pleasures of Science or intellectual Improvements are those it is most fond of, it has then a noble and refined Taste; but if its chief Satisfactions derive from Religion and Divine Contemplation, it has then its true and proper Taste; its Temper is as it should be, pure, divine, and heavenly; provided these Satisfactions spring from a true, religious Principle, free from that Superstition, Bigotry and Enthusiasm, under which it is often disguised.

And thus by carefully observing what it is that gives the Mind the greatest Pain and Torment, or the greatest Pleasure and Entertainment, we come at the Knowledge of its reigning Passions, and prevailing Temper and Disposition.

'Include thyself then, O my Soul, within
' the Compass of thine own Heart; if it be not
' large, it is deep; and thou wilt there find Ex-
' ercise enough. Thou wilt never be able to
' found it; it cannot be known, but by Him,
' who

'who tries the Thoughts and Reins. But dive
'into this Subject as deep as thou canst. Exa-
'mine thyself; and this Knowledge of that which
'passes within thee will be of more Use to thee,
'than the Knowledge of all that passes in the
'World. Concern not thyself with the Wars
'and Quarrels of publick or private Persons.
'Take Cognizance of those Contests which are
'between thy Flesh and thy Spirit; betwixt the
'Law of thy Members, and that of thy Under-
'standing. Appease those Differences. Teach
'thy Flesh to be in Subjection. Replace Reason
'on its Throne; and give it Piety for its Coun-
'sellor. Tame thy Passions, and bring them
'under Bondage. Put thy little State in good Or-
'der; govern wisely and holily those numerous
'People which are contained in so little a King-
'dom; that is to say, that Multitude of Affecti-
'ons, Thoughts, Opinions and Passions which
'are in thine Heart (*n*).'

C H A P. XI.

Concerning the secret Springs of our Actions.

X. ANOTHER *considerable Branch of Self-acquaintance is, to know the true Motives and secret Springs of our Actions.*

This

(*n*.) *Juricu's Method of Christian Devotion,* Part iii. Chap. iii.

This will sometimes cost us much Pains to acquire. But for want of it, we shall be in Danger of passing a false Judgment upon our Actions, and of entertaining a wrong Opinion of our Conduct.

It is not only very possible, but very common, for Men to be ignorant of the chief Inducements of their Behaviour; and to imagine they act from one Motive, whilst they are apparently governed by another. If we examine our Views, and look into our Hearts narrowly, we shall find that they more frequently deceive us in this respect than we are aware of; by persuading us that we are governed by much better Motives than we really are. The Honour of GOD, and the Interest of Religion, may be the open and avowed Motive; whilst secular Interest and secret Vanity may be the hidden and true one. While we think we are serving GOD, we may be only sacrificing to *Mammon*. We may, like *Jehu*, boast our *Zeal for the Lord*, when we are only animated by the Heat of our natural Passions (*o*); may cover a censorious Spirit under a Cloak of Piety; and giving Admonition to others, may be only giving Vent to our Spleen.

Many come to the Place of publick Worship, out of Custom or Curiosity, who would be thought to come thither only out of Conscience. And whilst their external and professed View is to serve GOD, and gain Good to their Souls, their secret

(*o*) 2 *Kings* x. 16.

secret and inward Motive is only to shew themselves to Advantage, or to avoid Singularity, and prevent others making Observations on their Absence. Munificence and Almsgiving may often proceed from a Principle of Pride and Party-Spirit, and seeming Acts of Friendship, from a mercenary Motive.

By thus disguising our Motives we may impose upon Men, but at the same time we impose upon ourselves; and whilst we are deceiving *others*, our own Hearts deceive *us*. And of all impostures *Self-deception* is the most dangerous, because least suspected.

Now, unless we examine this Point narrowly, we shall never come to the Bottom of it; and unless we come at the true Spring and real Motive of our Actions, we shall never be able to form a right Judgment of them; and they may appear very different in our own Eye, and in the Eye of the World, from what they do in the Eye of GOD. *For the LORD seeth not as Man seeth: for Man looketh on the outward Appearance, but the LORD looketh on the Heart* *. And hence it is, that *that which is highly esteemed among Men, is* oftentimes *Abomination in the Sight of GOD* †. *Every Way of Man is right in his own Eyes: but the LORD pondereth the Hearts* ‡.

CHAP.

---

* 1 Sam. xvi. 7.   † Luke xvi. 15.   ‡ Prov. xxi. 2.

CHAP. XII.

Every one that knows himself, is in a particular Manner sensible how far he is governed by a Thirst for Applause.

XI. ANOTHER *Thing necessary to unfold a Man's Heart to himself is, to consider what is his* Appetite for Fame; *and by what Means he seeks to gratify it.*

This Passion in particular having always so main a Stroke, and oftentimes so unsuspected an Influence on the most important Parts of our Conduct, a perfect Acquaintance with it is a very material Branch of *Self-Knowledge*, and therefore requires a distinct Consideration.

Emulation, like the other Passions of the Human Mind, shows itself much more plainly, and works much more strongly in some than it does in others. It is in itself innocent; and was planted in our Natures for very wise Ends, and, if kept under proper Regulations, is capable of serving very excellent Purposes, otherwise it degenerates into a mean and criminal *Ambition*.

When a Man finds something within him that pushes him on to excel in worthy Deeds, or in Actions truly good and virtuous, and pursues that Design with a steady unaffected Ardour, without Reserve or Falsehood, it is a true Sign of a noble Spirit. For that Love of Praise can never be criminal, that excites and enables a Man to do a great deal more Good than he could do without it. And perhaps there never was a fine Genius or a noble Spirit, that rose above the common Level, and distinguished itself by high Attainments in what is truly excellent, but was secretly, and perhaps insensibly, prompted by the Impulse of this Passion.

But, on the contrary, if a Man's Views centre only in the Applause of others, whether it be deserved or not; if he pants after Popularity and Fame, not regarding how he comes by it; if his Passion for Praise urge him to *stretch himself beyond the Line* of his Capacity, and to attempt Things to which he is unequal; to condescend to mean Arts and low Dissimulation for the Sake of a Name; and in a sinister, indirect Way, sue hard for a little Incense, not caring from whom he receives it; his Ambition then becomes Vanity. And if it excite a Man to wicked Attempts, make him willing to sacrifice the Esteem of all wise and good Men to the Acclamations of a Mob; to overleap the Bounds of Decency and Truth, and break through the Obligations of Honour and Virtue, it is then not only Vanity, but *Vice*; a Vice the most destructive to the Peace and Happiness of Human Society, and which of all others hath made the greatest Havock and Devastation among Men.

What an Instance have we here of the wide Difference between common Opinion and Truth? That a Vice so big with Mischief and Misery should be mistaken for a Virtue! And that they who have been most infamous for it should be crowned with Laurels, even by those who have been ruined by it; and have those Laurels perpetuated by the common Consent of Men through After ages! Seneca's Judgment of *Alexander* is certainly more agreeable to Truth than the common Opinion; who called him " a publick Cut-throat " rather than a Hero; and who, in seeking only " to be a Terror to Mankind, arose to no greater " an Excellence, than what belonged to the most " hurtful and hateful Animals on Earth (*p*)."

(*p*) Quid enim simile habebat vesanus Adolescens, cui pro virtute erat felix Temeritas? —— Hic a pueritiâ latro, gentiumque Vastator, tam hostium pernicies quam amicorum. Qui summum bonum duceret terori esse cunctis mortalibus: oblitus non ferocissima tantùm, sed ignavissima quoque animalia, timeri ob virus malum. *Sen. de Benef. cap.* 13.

How different from this is the Judgment of *Plutarch* in this Matter? who, in his *Oration concerning the Fortune and Virtue of Alexander*, exalts him into a *true Hero*; and justifies all the Waste he made of Mankind under (the same Colour with which the *Spaniards* excused their inhuman Barbarities towards the poor *Indians*, viz.) a Pretence of civilizing them. And in attributing all his Success to his Virtue, he talks more like a *Soldier* serving under him in his Wars, than an Historian who lived many Years afterwards, whose Business it was to transmit his Character impartially to future Ages. And in whatever other respects Mr. *Dryden* may give the Preference to *Plutarch* before *Seneca*, (which he does with much Zeal in his *Preface* to *Plutarch's Lives*) yet it must be allowed that, in this Instance at least, the latter shows more of the *Philosopher*. See *Plut. Mor. Vol.* i. *ad fin.*

Certain it is, that these false Heroes who seek their Glory from the Destruction of their own Species, are of all Men most ignorant of themselves; and by this wicked Ambition entail Infamy and Curses upon their Name, instead of that immortal Glory they pursued. According to the Prophet's Words, *Woe to him who coveteth an evil Coveteousness to his House, that he may set his Nest on high; that he may be delivered from the Power of Evil. Thou hast consulted Shame to thine House, by cutting off many People; and hast sinned against thy Soul* (q).

Now no Man can truly know himself till he be acquainted with this, which is so often the secret and unperceived Spring of his Actions, and observes how far it governs him in his Conversation and Conduct; Virtue and real Excellence will rise to View, tho' they be not mounted on the Wings of Ambition, which, by soaring too high, procures but a more fatal Fall.

And to correct the Irregularity and Extravagance of this Passion, let us but reflect how airy and unsubstantial a Pleasure the highest Gratifications of it afford; how many cruel Mortifications

---

(q) Hab. ii. 9, 10. בֹּצֵעַ בֶּצַע רָע: *that gaineth a wicked Gain.*

<blockquote>
Oh Sons of Earth! Attempt ye still to rise,
By Mountains pil'd on Mountains, to the Skies?
Heav'n still with Laughter the vain Toil surveys,
And buries Madmen in the Heaps they raise.
Who wickedly is wise, or madly brave,
Is but the more a Fool, or more a Knave.
</blockquote>

Pope's Essay on Man.

Chap. XII. *doth consist.* 101

tions it exposes us to, by awakening the Envy of others; to what Meanness it often makes us submit; how frequently it loseth its End by pursuing it with too much Ardor; and how much more solid Pleasure the Approbation of Conscience will yield, than the Acclamations of ignorant and mistaken Men, who, judging by Externals only, cannot know our true Character; and whose Commendations a wise Man would rather despise than court. ' Examine but the Size of People's
' Sense, and the Condition of their Understand-
' ing, and you will never be fond of Popu-
' larity, nor afraid of Censure; nor solicitous
' what Judgment they may form of you, who know
' not how to judge rightly of themselves (*r*).'

(*r*) Διελθε ισω εις τα ηγεμονικα αυτων, και οψει τινας κριτας φοβη οιος και περι αυτων οιας κρισεις. *Mark Anton. lib.* ix. § 18.

CHAP. XIII.

What kind of Knowledge we are already furnished with, and what Degree of Esteem we set upon it.

XII. A MAN *can never rightly know himself, unless he examines into his Knowledge of other Things.*

We must consider then the Knowledge we have; and whether we do not set too high a Price upon it, and too great a Value upon ourselves on the Account of it; of what real Use it is to us, and what Effect it has upon us; whether it does not make us too stiff, unsociable, and assuming; testy and supercilious, and ready to despise others for their supposed Ignorance. If so, our Knowledge, be it what it will, does us more Harm than Good. We were better without it; Ignorance itself would not render us so ridiculous. Such a Temper, with all our Knowledge, shows that we *know not ourselves*.

' A Man is certainly proud of that Knowledge
' he despises others for the Want of.'

How common is it for some Men to be fond of appearing to know more than they do, and of seeming to be thought Men of Knowledge? To which End they exhaust their Fund almost in all Companies, to outshine the rest. So that in two or three Conversations they are drawn dry, and you see to the Bottom of them much sooner than you could at first imagine. And even that Torrent of Learning, which they pour out upon you at first so unmercifully, rather confounds than satisfies you; their visible Aim is not to inform *your* Judgment, but display their *own*; you have many Things to query and except against, but their Loquacity gives you no room; and their good Sense, set off to so much Advantage, strikes a modest Man dumb: If you insist upon your Right to examine, they retreat, either in Confusion or Equivocation; and, like the Scuttle Fish, throw a large Quantity of Ink behind them, that you may not see where to pursue. Whence this Foible flows is obvious enough. *Self-Knowledge* would soon correct it.

But as some ignorantly affect to be more knowing, so others vainly affect to be more ignorant than they are; who, to shew they have greater Insight and Penetration than other Men, insist upon the absolute Uncertainty of Science; will dispute even first Principles; grant nothing as certain, and so run into downright *Pyrronism*; the too common Effect of abstracted Debates *excessively* refined (*s*).

(*s*) *Socrates*'s Saying. Nihil se scire, nisi id ipsum, favoured of an affected Humility. But they that followed went further; and particularly *Arcesilas*, Negabat esse quicquam, quod sciri potest; ne illud quidem ipsum quod Socrates sibi reliquisset. And thus the absurdity grew to a Size that was monstrous. *For to know that one knows nothing, is a Contradiction. And not to know that*

Every one is apt to set the greatest Value upon that Kind of Knowledge, in which he imagines he himself most excels; and to undervalue all other in Comparison of it. There wants some certain Rule then, by which every Man's Knowledge is to be tried, and the Value of it estimated. And let it be this.——"That is the best "and most valuable Kind of Knowledge, that "is most subservient to the best Ends; *i. e.* which "tends to make a Man wiser and better, or "more agreeable and useful both to himself and "others."———For Knowledge is but a *Means* that relates to some *End*. And as all Means are to be judged of by the Excellency of the End, and their Expediency to produce it; so *that* must be the best Knowledge that hath the *directest* Tendency to promote the *best* Ends; *viz.* a Man's own true Happiness, and that of others; in which the Glory of GOD, the ultimate End, is ever necessarily comprised.

Now, if we were to judge of the several Kinds of Science by this Rule, we should find, (1.) Some of them to be very *hurtful* and *pernicious*; as tending to pervert the true End of Knowledge; to ruin a Man's own Happiness, and make him more injurious to Society. Such is the Knowledge of Vice, the various Temptations to it, and the secret Ways of practising it; especially the Arts of Dissimulation, Fraud, and Dishonesty. (2.) Others

that he knows even that, is not to know but that he may know something. Relig. of Nat. delin. Pag. 40.

(2.) Others will be found *unprofitable* and *useless*. As those Parts of Knowledge, which tho' they may take up much Time and Pains to acquire, yet answer no valuable Purpose; and serve only for Amusement, and the Entertainment of the Imagination. For Instance, an Acquaintance with Plays, Novels, Games, and Modes, in which a Man may be very critical and expert, and yet not a whit the wiser or more useful Man. (3.) Other Kinds of Knowledge are good only *relatively*, or conditionally, and may be more useful to one than to another; *viz.* a Skill in a Man's particular Occupation or Calling, on which his Credit, Livelihood, or Usefulness in the World depends. And as this Kind of Knowledge is valuable in Proportion to its End, so it ought to be cultivated with a Diligence and Esteem answerable to that. (Lastly,) Other Kinds of Knowledge are good *absolutely* and universally; *viz.* the Knowledge of GOD and ourselves. The Nature of our final Happiness, and the Way to it. This is equally necessary to all. And how thankful should we be, that we, who live under the Light of the Gospel, and enjoy that Light in its Perfection and Purity, have so many happy Means and Opportunities of attaining this most useful and necessary Kind of Knowledge!

A Man can never understand himself then, till he makes a right Estimate of his Knowledge; till he examines what Kind of Knowledge he values himself most upon, and most diligently cultivates;

how high a Value he sets upon it; what Good it does him; what Effect it hath upon him; what he is the better for it; what End it answers now; or what it is like to answer hereafter.

There is nothing in which a Man's Self-ignorance discovers itself more, than in the Esteem he hath for his Understanding, or for himself on the Account of it. It is a trite and true Observation, that *empty Things make the most Sound.* Men of the least Knowledge are most apt to make a Show of it, and to value themselves upon it; which is very visible in forward confident Youth, raw conceited Academicks, and those who, uneducated in their Childhood, betake themselves in later Life to Reading, without Taste or Judgment, only as an Accomplishment, and to make a Show of Scholarship; who have just Learning enough to spoil Company, and render themselves ridiculous, but not enough to make either themselves or others at all the wiser.

But beside the forementioned Kinds of Knowledge, there is another which is commonly called *false Knowledge*; which tho' it often imposes upon Men under the Show and Semblance of true Knowledge, is really worse than Ignorance. Some Men have learned a great many Things, and have taken a great deal of Pains to learn them, and stand very high in their own Opinion on Account of them, which yet they must unlearn before they are truly wise. They have been at a vast Expence of Time, and Pains, and Patience,

ence, to heap together, and to confirm themselves in a Set of wrong Notions, which they lay up in their Minds as a Fund of valuable Knowledge; which, if they try by the forementioned Rules, *viz.* " the Tendency they have to *make them wiser and better,* or *more useful and beneficial to others,*" will be found to be worth just nothing at all.

Beware of this false Knowledge. For as there is nothing of which Men are more obstinately tenacious, so there is nothing that renders them more vain, or more averse to *Self-Knowledge.* Of all Things, Men are most fond of their wrong Notions.

The Apostle *Paul* often speaks of these Men, and their Self-sufficiency, in very poignant Terms; who, *tho' they seem wise,* yet (says he) *must become Fools before they are wise* *. Tho' they think they know a great deal, *know nothing yet as they ought to know* †. But *deceive themselves, by thinking themselves something when they are nothing* ‡. And whilst they *desire to be Teachers of others, understand not what they say, nor whereof they affirm* §. And *want themselves to be taught what are the first Rudiments and Principles* of Wisdom ‖.

* 1 Cor. iii. 18.   † 1 Cor. viii. 2.   ‡ Gal. vi. 3.
§ 1 Tim. i. 7.   ‖ Heb. v. 12.

CHAP

CHAP. XIV.

Concerning the Knowledge, Guard, and Government of our Thoughts.

XIII. ANOTHER *Part of Self-Knowledge consists in a due Acquaintance with our own Thoughts, and the inward Workings of the Imagination.*

The right Government of the *Thoughts* requires no small Art, Vigilance, and Resolution. But it is a Matter of such vast Importance to the Peace and Improvement of the Mind, that it is worth while to be at some Pains about it. A Man that hath so numerous and turbulent a Family to govern as his own Thoughts, which are too apt to be at the Command of his *Passions* and *Appetites*, ought not to be long from Home. If he be, they will soon grow mutinous and disorderly under the Conduct of those two head-strong Guides, and raise great Clamours and Disturbances, and sometimes on the slightest Occasions. And a more dreadful Scene of Misery can hardly be imagined, than that which is occasioned by such a Tumult and Uproar within, when a raging Conscience or inflamed Passions are let loose without Check or Controul. A City in Flames, or the Mutiny of a drunken Crew *aboard*, who have murdered the *Captain*, and are butchering one another, are but faint Emblems of it. The Torment

ment of the Mind, under such an Insurrection and Ravage of the Passions, is not easy to be conceived. The most revengeful Man cannot wish his enemy a greater.

Of what vast Importance then is it for a Man to watch over his Thoughts, in order to a right Government of them! To consider what Kind of Thoughts find the easiest Admission, in what Manner they insinuate themselves, and upon what Occasions!

It was an excellent Rule which a wise Heathen prescribed to himself, in his private Meditations; *Manage* (saith he) *all your Actions and Thoughts in such a Manner, as if you were just going out of the World* (t). Again, (saith he) *a Man is seldom, if ever, unhappy for not knowing the Thoughts of others; but he that does not attend to the Motions of his own, is certainly miserable* (u).

It may be worth our while then to discuss this Matter a little more precisely; and consider, (1.) *What kind of Thoughts are to be excluded or rejected.* And (2.) *What ought to be indulged and entertained.*

I. Some

(t) *Marc. Anton. Medit. lib.* 2. §. 11.
(u) *Marc. Anton. lib.* 2. §. 8.

" Nothing can be more unhappy than that Man, who ranges
" every where, ransacks every Thing, digs into the Bowels of the
" Earth, dives into other Men's Bosoms, but does not consider
" all the while that his own Mind will afford him sufficient Scope
" for Enquiry and Entertainment, and that the Care and Improve-
" ment of himself will give him Business enough. *Id. lib.* 2. §. 13.

" Your Disposition will be suitable to that which you most fre-
" quently think on; for the Soul is, as it were, tinged with the
" Colour and Complexion of its own Thoughts. *Id. lib.* 5. §. 16.

I. *Some Thoughts ought to be immediately banished as soon as they have found Entrance.*——And if we are often troubled with them, the safest Way will be to keep a good Guard on the Avenues of the Mind by which they enter, and avoid those Occasions which commonly excite them. For sometimes it is much easier to prevent a bad Thought entering the Mind, than to get rid of it when it is entered.——More particularly,

(1.) Watch against all *fretful* and *discontented* Thoughts, which do but chafe and corrode the Mind to no Purpose. To harbour these is to do yourself more Injury than it is in the Power of your greatest Enemy to do you. It is equally a Christian's Interest and Duty to *learn, in whatever State he is, therewith to be content* *.

(2.) Harbour not too *anxious* and *apprehensive* Thoughts. By giving Way to tormenting Fears, Suspicions of some approaching Danger or troublesome Event, we not only anticipate, but double the Evil we fear; and undergo much more from the Apprehension of it before it comes, than from the whole Weight of it when present. This is a great, but common Weakness; which a Man should endeavour to arm himself against by such kind of Reflections as these;——" Are not all
" these Events under the certain Direction of a
" wise Providence? If they befal me, they are
" then that Share of Suffering which GOD hath
" appointed me; and which he expects I should
" bear

* *Phil.* iv. 11.

"bear as a Christian. How often hath my too
"timorous Heart magnified former Trials? which
"I found to be less in Reality than they appeared
"upon their Approach. And perhaps the formi-
"dable Aspect they put on, is only a Stratagem
"of the great Enemy of my best Interest, design-
"ed on Purpose to divert me from some Point of
"Duty, or to draw me into some Sin, to avoid
"them. However, why should I torment myself
"to no Purpose? The Pain and Affliction the
"dreaded Evil will give me when it comes, is of
"GOD's sending; the Pain I feel in the Ap-
"prehension of it before it comes, is of my own
"procuring. Whereby I often make my Suf-
"ferings more than double; for this Overplus of
"them, which I bring upon myself, is often great-
"er than that Measure of them which the Hand
"of Providence immediately brings upon me."

(3.) Dismiss, as soon as may be, all *angry* and *wrathful* Thoughts. These will but canker and corrode the Mind, and dispose it to the worst Temper in the World, *viz.* that of fixed *Malice* and *Revenge*. Anger may steal into the Heart of a wise Man, but it *rests* only *in the Bosom of Fools* *. Make all the most candid Allowances for the Offender. Consider his natural Temper. Turn your Anger into Pity. Repeat 1 *Cor.* xiii. Think of the Patience and Meekness of *Christ*, and the Petition in the *Lord's-Prayer*; and how much you stand in need of Forgiveness yourself,
both

* *Eccles.* vii. 9.

both from GOD and Man; how fruitless, how foolish is indulged Resentment; how tormenting to yourself? You have too much Good-Nature willingly to give others so much Torment; and why should you give it yourself? You are commanded to *love your Neighbour as yourself*, but not forbidden to love yourself as much. And why should you do yourself that Injury, which your Enemy would be glad to do you * ?

But, above all, be sure to set a Guard on the Tongue, whilst the fretful Mood is upon you. The least Spark may break out into a Conflagration, when cherished by a resentive Heart, and fanned by the Wind of an angry Breath. Aggravating Expressions at such a Time, are like Oil thrown upon Flames, which always make them rage the more (*w*). Especially,

(4.) Banish all *malignant* and *revengeful* Thoughts. A Spirit of Revenge is the very Spirit of the Devil; than which nothing makes a Man more like him; and nothing can be more opposite to the Temper which Christianity was designed to promote. If your Revenge be not satisfied, it

---

* The Christian Precept in this Case is, *let not the Sun go down upon your Wrath*, Eph. iv. 26. And this precept *Plutarch* tells us the *Pythagoreans* practised in a literal Sense: "Who, if at any "Time in a Passion they broke out into opprobrious Language, be- "fore Sun set gave one another their Hands, and with them a "Discharge from all Injuries; and so with a mutual Reconcilia- "tion parted Friends." *Plut. Mor. Vol.* iii. *Pag.* 89.

(*w*) Αγαθον μεν εςι εν πυρετω, δε εν οργη την γλωτταν απαλην εχειν και λειαν. Plutarch. *de Irá Cobiben.* It is good in a Fever, much better in Anger, to have the Tongue kept clean and smooth.

it will give you Torment *now*; if it be, it will give you greater *hereafter.* None is a greater Self-Tormentor than a malicious and revengeful Man, who turns the Poison of his own Temper in upon himself (*x*).

(5.) Drive from the Mind all *silly, trifling,* and *unreasonable* Thoughts; which sometimes get into it we know not how, and seize and possess it before we are aware; and hold it in empty, idle Amusements, that yield it neither Pleasure nor Profit, and turn to no manner of Account in the World; only consume Time, and prevent a better Employment of the Mind. And indeed there is little Difference whether we spend the Time in Sleep, or in these waking Dreams. Nay, if the Thoughts which thus insensibly steal upon you be not altogether absurd and whimsical, yet if they be impertinent and unseasonable, they ought to be dismissed, because they keep out better Company.

(6.) Cast out all *wild* and *extravagant* Thoughts, all *vain* and *fantastical* Imaginations. Suffer not your Thoughts to roam upon Things that never were, and perhaps never will be; to give you a visionary Pleasure in the Prospect of what you have not the least Reason to hope, or a needless Pain in the Apprehension of what you have not the least Reason to fear.——The Truth is, next to

---

(*x*) Malitia ipsa maximam partem veneni sui bibit.—— Illud venenum quod serpentes in alienam perniciem proferunt, sine suâ continent. Non est huic simile; hoc habentibus pessimum est. *Sen. Epist.* 82.

to a clear Conscience and a sound Judgment, there is not a greater Blessing than a *regular* and *well-governed Imagination*; to be able to view Things as they are, in their true Light and proper Colours; and to distinguish the false Images that are painted on the Fancy, from the Representations of Truth and Reason. For how common a Thing is it for Men, before they are aware, to confound Reason and Fancy, Truth and Imagination together? To take the Flashes of the animal Spirits for the Light of Evidence? and think they believe Things to be true or false, when they only *fancy* them to be so? and fancy them to be so, because they *would have* them so? Not considering that mere Fancy is only the *Ignis fatuus* of the Mind; which often appears brightest, when the Mind is most covered with Darkness; and will be sure to lead them astray, who follow it as their Guide. Near akin to these are,

(7.) *Romantick* and *chimerical* Thoughts. By which I mean that Kind of Wild-fire, which the Briskness of the animal Spirits sometimes suddenly flashes on the Mind, and excites Images that are so *extremely* ridiculous and absurd, that one can scarce forbear wondering how they could get Admittance. These random Flights of the Fancy are soon gone; and herein differ from that Castle-building of the Imagination before-mentioned, which is a more settled Amusement. But these are too incoherent and senseless to be of long Continuance; and are the maddest Sallies, and the

most

most ramping Reveries of the Fancy that can be. —— I know not whether my Reader understands *now* what I mean; but if he attentively regards all that passes through his Mind, perhaps he may hereafter by Experience.

(8.) Repel all *impure* and *lascivious* Thoughts; which taint and pollute the Mind; and tho' hid from Men, are known to GOD, in whose Eye they are abominable. Our Saviour warns us against these as a kind of Spiritual *Fornication*[*], and inconsistent with that *Purity of Heart* which his Gospel requires.

(9.) Take care how you too much indulge *gloomy* and *melancholy* Thoughts. Some are disposed to see every Thing in the worst Light. A black Cloud hangs hovering over their Minds; which, when it falls in Showers through their Eyes, is dispersed; and all within is serene again. This is often purely mechanical; and owing either to some Fault in the bodily Constitution, or some accidental Disorder in the animal Frame. However, one that consults the Peace of his own Mind will be upon his Guard against this, which so often robs him of it.

(10.) On the other Hand, let not the Imagination be too *sprightly* and *triumphant*. Some are as unreasonably exalted, as others are depressed; and the same Person at different Times often runs into both Extremes; according to the different Temper and Flow of the animal Spirits.
And

[*] *Matt.* v. 28.

And therefore the Thoughts, which so eagerly crowd into the Mind at such Times, ought to be suspected and well guarded; otherwise they will impose upon our Judgments, and lead us to form such a Notion of ourselves and of Things, as we shall soon see fit to alter, when the Mind is in a more settled and sedate Frame.

Before we let our Thoughts judge of Things, we must set Reason to judge our Thoughts; for they are not always in a proper Condition to execute that Office. We do not believe the Character which a Man gives us of another, unless we have a good Opinion of his own; so neither should we believe the Verdict which the Mind pronounces, till we first examine whether it be impartial and unbiassed; whether it be in a proper Temper to judge, and have proper Lights to judge by. The Want of this previous Act of *Self-judgment*, is the Cause of much Self-deception and false Judgment.

(Lastly,) With Abhorrence reject immediately all *profane* and *blasphemous* Thoughts; which are sometimes suddenly injected into the Mind, we know not *how*, tho' we may give a pretty good Guess *from whence*. And all those Thoughts which are apparently Temptations and Inducements to Sin, our Lord hath, by his Example, taught us to treat in this Manner *.

These then are the Thoughts we should carefully guard against.———And as they will (especially some of them) be frequently insinuating them-

* *Matt.* iv. 10.

themselves into the Heart, remember to set Reason at the Door of it to guard the Passage, and bar their Entrance, or drive them out forthwith when entered; not only as impertinent, but mischievous Intruders.

But, II. There are other Kinds of Thoughts which we ought to *indulge*, and with great Care and Diligence *retain* and *improve*.

Whatever Thoughts give the Mind a rational or religious Pleasure, and tend to improve the Heart and Understanding, are to be favoured, often recalled, and carefully cultivated. Nor should we dismiss them, till they have made some Impressions on the Mind, which are like to abide there.

And to bring the Mind into a Habit of recovering, retaining, and improving such Thoughts, two Things are necessary.

(1.) To habituate ourselves to a *close* and *rational Way of thinking*. And, (2.) To *moral Reflections* and *religious Contemplations*.

(1.) To prepare and dispose the Mind for the Entertainment of good and useful Thoughts, we must take care to accustom it to a *close* and *rational* Way of Thinking.

When you have started a good Thought, pursue it; do not presently lose Sight of it, or suffer any trifling Suggestion that may intervene to divert you from it. Dismiss it not till you have sifted and exhausted it; and well considered the several Consequences and Inferences that result from

from it. However, retain not the Subject any longer than you find your Thoughts run freely upon it; for to confine them to it when it is quite worn out, is to give them an unnatural Bent, without sufficient Employment; which will make them flag, or be more apt to run off to something else.

And to keep the Mind intent on the Subject you think of, you must be at some Pains to recal and refix your defultory and rambling Thoughts. Lay open the Subject in as many Lights and Views as it is capable of being reprefented in. Clothe your beft Ideas in pertinent and well-chofen Words, deliberately pronounced; or commit them to Writing.

Whatever be the Subject, admit of no Inferences from it, but what you fee plain and natural. This is the Way to furnifh the Mind with true and folid Knowledge. As on the contrary, *falfe* Knowledge proceeds from not underftanding the Subject, or drawing Inferences from it which are forced and unnatural; and allowing to thofe precarious Inferences, or Confequences drawn from *them*, the fame Degree of Credibility as to the moft rational and *beft eftablifhed* Principles.

Beware of a *fuperficial*, *flight*, or *confufed* View of Things. Go to the Bottom of them, and examine the Foundation; and be fatisfied with none but clear and diftinct Ideas (when they can be had) in every Thing you read, hear, or think of.

of. For resting in imperfect and obscure Ideas, is the Source of much Confusion and Mistake.

Accustom yourself to *speak* naturally, pertinently, and rationally, on all Subjects, and you will soon learn to *think* so on the best; especially if you often converse with those Persons that speak, and those Authors that write, in that Manner.

Such a Regulation and right Management of your Thoughts and rational Powers, will be of great and general Advantage to you, in the Pursuit of youthful Knowledge, and a good Guard against the Levities and frantick Sallies of the Imagination. Nor will you be sensible of any Disadvantage attending it, excepting one, *viz.* its making you more sensible of the Weakness and Ignorance of others, who are often talking in a random, inconsequential Manner; and whom it may oftentimes be more prudent to bear with, than contradict. But the vast Benefit this Method will be of in tracing out Truth and detecting Error, and the Satisfaction it will give you in the cool and regular Exercises of Self-employment, and in the retaining, pursuing, and improving good and useful Thoughts, will more than compensate that petty Disadvantage.

(2.) If we would have the Mind furnished and entertained with good Thoughts, *we must inure it to moral and religious Subjects.*

It is certain the Mind cannot be more nobly and usefully employed than in such Kind of Contemplations. Because the Knowledge it thereby acquires,

acquires, is of all others the most excellent Knowledge; and that both in regard of its *Object* and its *End*; the Object of it being GOD, and the End of it *Eternal Happiness*.

The great End of Religion is to *make us like GOD*, and *conduct us to the Enjoyment of Him*. And whatever hath not this plain Tendency, and especially if it have the contrary, Men may call *Religion* (if they please) but they cannot call it more out of its Name. And whatever is called religious Knowledge, if it does not direct us in the Way to this End, is not religious Knowledge; but something else *falsely so called*. And some are unhappily accustomed to such an Abuse of Words and Understanding, as not only to call, but to *think* those Things *Religion*, which are quite the reverse of it; and those Notions *religious Knowledge*, which lead them the farthest from it.

The Sincerity of a true religious Principle cannot be better known, than by the Readiness with which the Thoughts advert to GOD, and the Pleasure with which they are employed in devout Exercises. And though a Person may not always be so well pleased with hearing religious Things talked of by others, whose different Taste, Sentiments, or Manner of Expression may have something disagreeable; yet if he have no Inclination to think of them himself, or to converse with himself about them, he hath great Reason to suspect that his *Heart is not right with GOD*. But if he frequently and delightfully exercise

ercise his Mind in divine Contemplations, it will not only be a good Mark of his Sincerity, but will habitually dispose it for the Reception of the best and most useful Thoughts, and fit it for the noblest Entertainments.

Upon the whole then, it is of as great Importance for a Man to take heed what Thoughts he entertains, as what Company he keeps; for they have the same Effect upon the Mind. Bad Thoughts are as infectious as bad Company; and good Thoughts solace, instruct, and entertain the Mind, like good Company. And this is one great Advantage of Retirement; that a Man may chuse what Company he pleases from within himself.

As in the World we oftener light into bad Company than good, so in Solitude we are oftener troubled with impertinent and unprofitable Thoughts, than entertained with agreeable and useful Ones. And a Man that hath so far lost the Command of himself, as to lie at the Mercy of every foolish or vexing Thought, is much in the same Situation as a *Host*, whose House is open to all *Comers*; whom, tho' ever so noisy, rude, and troublesome, he cannot get rid of; but with this Difference, that the *latter* hath some Recompence for his Trouble, the *former* none at all; but is robbed of his Peace and Quiet for nothing.

Of such vast Importance to the Peace, as well as the Improvement of the Mind, is the right Regulation of the Thoughts. Which will be my Apology for dwelling so long on this Branch of the Subject; which I shall conclude with this one Observation more; that it is a very dangerous Thing to think, as too many are apt to do, that it is a Matter of Indifference what Thoughts they entertain in their Hearts; since the Reason of Things concurs with the Testimony of the Holy Scriptures to assure us, *that the* allowed *Thought of Foolishness is Sin* * (x).

CHAP. XV.

Concerning the Memory.

XIV. A MAN *that knows himself will have a Regard not only to the Management of his* Thoughts, *but the Improvement of his* Memory.

The Memory is that Faculty of the Soul, which was designed for the Store-house or Repository of its most useful Notions; where they may be laid up in Safety, to be produced upon proper Occasions.

Now a thorough Self-acquaintance cannot be had without a proper Regard to this in two Respects. (1.) Its Furniture. (2.) Its Improvement.

(1.) A Man that knows himself will have a Regard to the *Furniture* of his Memory; not to
load

---

* *Prov.* xxiv. 9.
(x) Nam scelus inter se tacitum qui cogitat ullum
   Facti crimen habet.   *Juv. Sat.* 13.
Guard well thy Thoughts; our Thoughts are heard in Heav'n.
   Young.

load it with Trash and Lumber, a Set of useless Notions or low Conceits, which he will be ashamed to produce before Persons of Taste and Judgment.

If the Retention be bad, do not crowd it. It is of as ill Consequence to overload a weak Memory, as a weak Stomach. And that it may not be cumbered with Trash, take heed what Company you keep, what Books you read, and what Thoughts you favour; otherwise a great deal of useless Rubbish may fix there before you are aware, and take up the Room which ought to be possessed by better Notions. But let not a valuable Thought slip from you, though you pursue it with much Time and Pains before you overtake it. The regaining and refixing it may be of more Avail to you than many Hours Reading.

What Pity it is that Men should take such immense Pains, as some do, to learn those Things which, as soon as they become wise, they must take as much Pains to *unlearn!* —— A Thought that should make us very curious and cautious about the proper Furniture of our Minds.

(2.) Self-Knowledge will acquaint a Man with the Extent and Capacity of his Memory, and the right Way to *improve* it (*y*).

(*y*) Tribus rebus potissimum constat optima memoria, Intellectu, Ordine, Cura, siquidem bona memoriæ pars est rem penitus *intellexisse*; tum *Ordo* facit, ut quæ semel exciderent, quasi postliminio in animum revocamus ; porro *Cura* omnibus in rebus, non hic tantum plurimum valet. *Erasm. de rat. stud. ad calc. Ringelbergii, p.* 168.

There is no small Art in improving a weak Memory, so as to turn it to as great an Advantage as many do theirs which are much stronger. A few short Rules to this Purpose may be no unprofitable Digression.

(1.) Beware of every Sort of *Intemperance* in the Indulgence of the Appetites and Passions. Excesses of all Kinds do a great Injury to the Memory.

(2.) If it be weak, do not *overlade* it. Charge it only with the most useful and solid Notions. A small Vessel should not be stuffed with Lumber. But if its Freight be precious, and judiciously stowed, it may be more valuable than a Ship of twice its Burthen.

(3.) Recur to the Help of a *Common-Place-Book*, according to Mr. *Locke's* Method; and review it once a Year. But take care that by confiding to your Minutes or memorial Aids, you do not excuse the Labour of the Memory; which is one Disadvantage attending this Method.

(4.) Take every Opportunity of uttering your best Thoughts in *Conversation*, when the Subject will admit it; that will deeply imprint them. Hence the Tales which common Story-tellers relate, they never forget, tho' ever so silly (z).

(5.) Join

(z) *Quicquid didiceris id confestim doceas; sic et tua firmare, et prodesse aliis potes.* Ringelbergius de ratione studii, p. 28.

Postremo illud nan ad unum aliquid, sed ad omnia simul plurimum conducet, si frequenter alios quoque doceas. Nusquam enim melius deprehenderis quid intelliges, quid non. Atque interim nova quæque occurrunt,

(5.) Join to the Idea you would remember some other that is more familiar to you, which bears some *Similitude* to it, either in its Nature, or in the Sound of the Word by which it is expressed; or that hath some Relation to it either in Time or Place. And then by recalling *this*, which is easily remembered, you will (by that *Concatenation*, or *Connection* of *Ideas* which Mr. *Locke* takes Notice of) draw in *that* which is thus linked or joined with it; which otherwise you might hunt after in vain.——This Rule is of excellent Use to help you to remember *Names*.

(6.) What you are determined to remember, think of before you go to sleep at *Night*, and the first Thing in the *Morning*, when the Faculties are fresh. And recollect at Evening every Thing worth remembring the Day past.

(7.) Think it not enough to furnish this Storehouse of the Mind with good Thoughts, but lay them up there in *Order*, digested or ranged under proper Subjects or Classes; that whatever Subject you have Occasion to think or talk upon, you may have Recourse immediately to a good Thought which you heretofore laid up there under *that Subject*. So that the very mention of the Subject may bring the Thought to Hand; by which Means you will carry a regular *Common-Place-book* in your Memory. And it may not be amiss sometimes to take an *Inventory* of this men-

tal

currunt, commentanti differentique, nihil non altius infigitur animo. Erasm. Rot. de rat. stud. *p.* 170.

tal Furniture, and recollect how many good Thoughts you have treasured up under such particular Subjects, and whence you had them.

(Lastly,) Nothing helps the Memory more than often *Thinking*, *Writing* or *Talking* on those Subjects you would remember.———But enough of this.

CHAP. XVI.

Concerning the Mental Taste.

XV. A MAN *that knows himself, is sensible of, and attentive to the particular* Taste *of his Mind, especially in Matters of Religion.*

As the late Mr. *Howe* judiciously observes, 'there is beside bare Understanding and Judg-
' ment, and diverse from that heavenly Gift,
' which in the Scripture is called Grace, such a
' Thing as *Gust* and *Relish* belonging to the Mind
' of Man, (and, I doubt not, with all Men, if
' they observe themselves) and which are as un-
' accountable, and as various as the Relishes and
' Disgusts of *Sense*. This *they* only wonder at
' who *understand not themselves*, or will consider
' no body but themselves.——— So that it cannot
' be said universally, that it is a better Judgment,
' or more Grace that determines Men the one
' Way or the other; but somewhat in the Tem-
' per of their Minds distinct from *both*, which
' I know not how better to express than by
' MEN-

'MENTAL TASTE.——— And this hath no
'more of Myſtery in it, than that there is ſuch a
'Thing belonging to our Natures as Complacen-
'cy and Diſpliecncy in reference to the Objects
'of the Mind. And this, in the Kind of it, is as
'common to Men as Human Nature; but as much
'diverſified in Individuals as Men's other Inclina-
'tions are'(a).

Now this different *Taſte* in Matters relating to Religion, (tho' it may be ſometimes natural or what is born with a Man, yet) generally ariſes from the Difference of Education and Cuſtom. And the true Reaſon why ſome Perſons have an inveterate Diſreliſh to certain Circumſtantials of Religion, tho' ever ſo juſtifiable, and at the ſame Time a fixed Eſteem for others that are more exceptionable, may be no better than what I have heard ſome very honeſtly profeſs, *viz.* that the one they have been uſed to, and the other not. As a Perſon by long Uſe and Habit acquires a greater Reliſh for coarſe and unwholeſome Food than the moſt delicate Diet; ſo a Perſon long habituated to a ſet of Phraſes, Notions, and Modes, may, by Degrees, come to have ſuch a Veneration and Eſteem for them, as to deſpiſe and condemn others which they have not been accuſtomed to, tho' perhaps more edifying, and more agreeable to Scripture and Reaſon.

This particular *Taſte* in Matters of Religion differs very much (as Mr. *Howe* well obſerves) both from Judgment and Grace.

How-

(a) See his *humble Requeſt both to Conformiſts and Diſſenters.*

However, it is often mistaken for both: When it is mistaken for the former, it leads to *Error*; when mistaken for the latter, to *Censoriousness*.

This different Taste of mental Objects is much the same with that, which with regard to the Objects of Sense we call *Fancy*; for as one Man cannot be said to have a better Judgment in Food than another, purely because he likes some Kind of Meats better than he; so neither can he be said to have a better Judgment in Matters of Religion, purely because he hath a greater Fondness for some particular Doctrines and Forms.

But though this *mental Taste* be not the same as the Judgment, yet it often draws the Judgment to it; and sometimes very much perverts it.

This appears in nothing more evidently than in the Judgment People pass upon the Sermons they hear. Some are best pleased with those Discourses that are *pathetic* and *warming*, others with what is more *solid* and *rational*, and others with the *sublime* and *mystical*; nothing can be too *plain* for the Taste of some, or too *refined* for that of others. Some are for having the Address only to their Reason and Understanding, others only to their Affections and Passions, and others to their Experience and Consciences. And every Hearer or Reader is apt to judge according to his particular Taste, and to esteem him the best Preacher or Writer who pleases him most; without examining first his own particular Taste, by which he judgeth.

It

It is natural indeed for every one to desire to have his own Taste pleased, but it is unreasonable in him to set it up as the best, and make it a Test and Standard to others. But much more unreasonable to expect that he who speaks in Publick should always speak to *his* Taste; which might as reasonably be expected by another of a different one. It is equally impossible that what is delivered to a Multitude of Hearers should alike suit all their Tastes, as that a single Dish, though prepared with ever so much Art and Exactness, should equally please a great Variety of Appetites; among which there may be some, perhaps, very nice and sickly.

It is the Preacher's Duty to adapt his Subjects to the Taste of his Hearers, as far as Fidelity and Conscience will admit; because it is well known from Reason and Experience, as well as from the Advice and Practice of the Apostle *Paul* (b), that this is the best Way to promote their Edification. But if their Taste be totally vitiated, and incline them to take in that which will do them more Harm than Good, and to relish Poison more than Food, the most charitable Thing the Preacher can do in that Case is, to endeavour to correct so vicious an Appetite, which loaths that which is most wholesome, and craves pernicious Food; this, I say, it is his Duty to attempt

---

(b) Rom. xv. 2. *Let every one of us please his Neighbour for his Good to Edification.* 1 Cor. ix. 22. *To the Weak, became I as weak, that I might gain the Weak: I am made all Things to all Men, that I might by all Means save some.*

tempt in the most gentle and prudent Manner he can, though he run the Risk of having his Judgment or Orthodoxy called into Question by them, as it very possibly may; for commonly they are the most arbitrary and unmerciful Judges in this Case, who are least of all qualified for that Office.

There is not perhaps a more unaccountable Weakness in Human Nature than this, that with regard to religious Matters our Animosities are generally greatest where our Differences are least; they who come pretty near to our Standard, but stop short there, are more the Objects of our Disgust and Censure, than they who continue at the greatest Distance from it. And in some Cases it requires much Candour and Self-command to get over this Weakness. To whatever secret Spring in the Human Mind it may be owing, I shall not stay to enquire; but the Thing itself is too obvious not to be taken Notice of.

Now we should all of us be careful to find out, and examine our proper *Taste* of religious Things; that if it be a false one, we may rectify it; if a bad one, mend it; if a right and good one, strengthen and improve it. For the Mind is capable of a false Gust, as well as the Palate; and comes by it the same Way; *viz.* by being long used to unnatural Relishes, which by Custom become grateful. And having found out what it is, and examined it by the Test of Scripture, Reason, and Conscience, if it be not very wrong let us indulge it, and read those Books that are most suited to it,

which

which for that Reason will be most edifying. But at the same time let us take Care of two Things, (1.) That it do not bias our Judgment, and draw us into *Error*. (2.) That it do not cramp our Charity, and lead us to *Censoriousness*.

C H A P. XVII.

Of our great and governing Views in Life.

XVI. ANOTHER *Part of* Self-Knowledge *is, to know what are the great Ends for which we live.*

We must consider what is the ultimate Scope we drive at; the general Maxims and Principles we live by; or whether we have not yet determined our End, and are governed by no fixed Principles; or by such as we are ashamed to own.

' The first and leading Dictate of Prudence is,
' that a Man propose to himself his true and best
' Interest for his End; and the next is that he
' make use of all those Means and Opportunities
' whereby that End is to be obtained. This is
' the most effectual Way that I know of to secure
' to one's self the Character of a wise Man here,
' and the Reward of one hereafter. And between
' these two there is such a close Connection, that he
' who does not do the *latter* cannot be supposed
' to intend the *former*. He that is not careful
' of his Actions shall never persuade me that he
' seriously proposes to himself his best Interest,
' as

' as his *End*; for if he did, he would as seriously
' apply himself to the Regulation of the other, as
' the *Means*' (c).

There are few that live so much at Random as not to have some main End in Eye; something that influences their Conduct, and is the great Object of their Pursuit and Hope. A Man cannot live without some *leading* Views; a wise Man will always know what they are, whether it is fit he should be led by them or no; whether they be such as his Understanding and Reason approve, or only such as Fancy and Inclination suggest. He will be as much concerned to *act* with Reason, as to talk with Reason; as much ashamed of a Solecism and Contradiction in his Character, as in his Conversation.

Where do our Views center? In this World we are in; or that we are going to? If our Hopes and Joys center here, it is a mortifying Thought, that we are every Day *departing from our Happiness*; but if they are fixed above, it is a Joy to think that we are every Day *drawing nearer* to the Object of our highest Wishes.

Is our main Care to appear great in the Eye of Man; or good in the Eye of GOD? If the former, we expose ourselves to the Pain of a perpetual Disappointment. For it is much if the Envy of Men do not rob us of a good deal of our just Praise, or if our Vanity will be content with that Portion of it they allow us. But if the latter

be

(c) *Norris's Misc. p.* 18.

be our main Care, if our chief View is to be approved of GOD, we are laying up a Fund of the moſt laſting and ſolid Satisfactions. Not to ſay that this is the trueſt Way to appear great in the Eye of Men; and to conciliate the Eſteem of all thoſe whoſe Praiſe is worth our Wiſh.

"Be this then, O my Soul, thy wiſe and ſteady
"Purſuit; let this circumſcribe and direct thy
"Views; be this a Law to thee, from which ac-
"count it a Sin to depart, whatever Diſreſpect or
"Contempt it may expoſe thee to from others (*d*);
"be this the Character thou reſolveſt to live up
"to, and at all Times to maintain both in pub-
"lick and private (*e*), *viz.* a Friend and Lover
"of GOD; in whoſe Favour thou centereſt all
"thy preſent and future Hopes. Carry this View
"with thee through Life, and dare not in any
"Inſtance to act inconſiſtently with it."

(*d*) Ὅσα προτίθεσαι, τούτοις ὡς νόμοις, καὶ ὡς ἀσεβησων ἂν παραβῆς τι τούτων ἐμμένε. Ὅτι δ' ἂν ἔρη τις περὶ σοῦ μὴ ἐπιστρέφου. *Epict. Ench. cap.* 74.——*What you have once wiſely propoſed ſtick to, as a Law not to be violated without Guilt. And mind not what others ſay of you.*

(*e*) Τάξον τινὰ ἤδη χαρακτῆρα σεαυτῷ, καὶ τύπον, ὃν φυλάξεις ἐπί τε σεαυτοῦ, καὶ ἀνθρώποις ἐντυχαίνων. *Idem. cap.* 40.——*Fix your Character, and keep to it; whether alone or in Company.*

CHAP.

CHAP. XVIII.

How to know the true State of our Souls; and whether we are fit to die.

LASTLY, the most important Point of Self-Knowledge, after all, is to know the true State of our Souls towards GOD; and in what Condition we are to die.

These two Things are inseparably connected in their Nature, and therefore I put them together. The Knowledge of the former will determine the latter, and is the only Thing that can determine it; for no Man can tell whether he is fit for Death, till he is acquainted with the true State of his own Soul.

This now is a Matter of such vast Moment, that it is amazing any considerate Man, or any one who thinks what it is to die, can be satisfied, so long as it remains an Uncertainty.——Let us trace out this important Point then with all possible Plainness; and see if we cannot come to some Satisfaction in it upon the most solid Principles.

In order to know then whether we are fit to die, we must first know *what it is that fits us for Death.*——And the Answer to this is very natural and easy; viz. that only fits us for Death, *that fits us for Happiness after Death.*

This is certain.——But the Question returns. *What is it that fits us for Happiness after Death?* Now

Now in Answer to this, there is a previous Question necessary to be determined; *viz. What that Happiness is?*

It is not a Fool's Paradise, or a Turkish Dream of sensitive Gratifications. It must be a Happiness suited to the Nature of the Soul, and what it is capable of enjoying in a State of Separation from the Body. And what can that be, but *the Enjoyment of* GOD, the best of Beings, and the Author of ours?

The Question then comes to this; *What is that which fits us for the* Enjoyment of GOD, in the future State of separate Spirits?

And methinks we may bring this Matter to a very sure and short Issue; by saying it is *that which makes us like to him now.* —— This only is our proper Qualification for the Enjoyment of him after Death, and therefore our only proper Preparation for Death. For how can they, who are unlike to GOD here, expect to enjoy him hereafter? And if they have no just Ground to hope that they shall enjoy GOD in the other World, how are they fit to die?——

So that the great Question, *Am I fit to die?* resolves itself into this, *Am I like to* GOD? For it is this only that fits me for Heaven; and that which fits me for Heaven, is the only thing that fits me for Death.

Let this Point then be well searched into, and examined very deliberately and impartially.

Most

Most certain it is, that GOD can take no real Complacency in any but those that are like Him; and it is as certain, that none but those that are like Him can take Pleasure in *Him*.—— But GOD is a most pure and holy Being; a Being of infinite Love, Mercy and Patience; whose Righteousness is invariable, whose Veracity inviolable, and whose Wisdom unerring. These are the moral Attributes of the Divine Being, in which he requires us to imitate Him; the express Lineaments of the Divine Nature, in which all good Men bear a Resemblance to Him; and for the Sake of which only they are the Objects of his Delight: For GOD can love none but those that bear this Impress of his own Image on their Souls. —— Do we find then these visible Traces of the Divine Image there? Can we make out our Likeness to him in his Holiness, Goodness, Mercy, Righteousness, Truth, and Wisdom? If so, it is certain we are capable of enjoying Him, and are the proper Objects of his Love.—— By this we know we are fit to die; because by this we know we are fit for Happiness after Death.

Thus then, if we are faithful to our Consciences, and impartial in the Examination of our Lives and Tempers, we may soon come to a right Determination of this important Question, *What is the true State of our Souls towards GOD? and in what Condition we are to die* *? Which as it is the most impor-

---

* " Nor do I apprehend the Knowledge of our State (call it Assurance if you please) so uncommon and extraordinary a Thing

important, so it is the last Instance of *Self-Knowledge* I shall mention: And with it close the first Part of this Subject.

" as some are apt to imagine. Understand by Assurance a satisfac-
" tory Evidence of the Thing, such as excludes all reasonable Doubts
" and disquieting Fears of the contrary, tho', it may be, not all
" transient Suspicions and Jealousies. And such an Assurance and
" Certainty Multitudes have attained, and enjoy the Comfort of;
" and indeed it is of so high Importance, that it is a Wonder any
" thoughtful Christian that believes an Eternity, can be easy one
" Week or Day without it." *Bennet's Christ. Orat. pag.* 569.

A TREATISE OF SELF-KNOWLEDGE.

PART II.

Shewing the great Excellency and Advantages of this Kind of Science.

HAVING in the former Part of the Subject laid open some of the main Branches of *Self-Knowledge*, or pointed out the principal Things which a Man ought to be acquainted with, relating to himself; I am now (Reader) to lay before you the Excellency and Usefulness of this Kind of Knowledge, (as an Inducement to labour after it) by a Detail of the several great Advantages attending it, which shall be recounted in the following Chapters.

CHAP.

CHAP. I.

Self-Knowledge the Spring of Self-possession.

I. ONE great *Advantage* of *Self-Knowledge* is, that it gives a Man the truest and most constant Self-possession.

A Man that is endowed with this excellent Knowledge is calm and easy,

(1.) Under *Affronts* and *Defamation*. For he thinks thus: 'I am sure I know myself better
' than any Man can pretend to know me. This
' Calumniator hath, indeed, at this Time missed
' his Mark, and shot his Arrows at random; and
' it is my Comfort, that my Conscience ac-
' quits me of his angry Imputation. However,
' there are worse Crimes which he might more
' justly accuse me of; which, though hid from
' him, are known to myself. Let me set about
' reforming them; lest, if they come to his No-
' tice, he should attack me in a more defenceless
' Part, find something to fasten his Oblo-
' quy, and fix a lasting Reproach upon my Cha-
' racter' (*f*).

There

(*f*) Εαν τις σοι απαγγειλη, ότι ο δεινα σε κακως λεγει, μη απολογε προς τα λεχθεντα· αλλ' αποκριν, ότι ηγνοει γαρ τα αλλα προσοντα μοι κακα, επει εκ αν ταυτα μονα ελεγεν. *Epict. Ench. cap.* 48.———*If you are told that another reviles you, do not go about to vindicate yourself, but reply thus; My other Faults I find are hid from him, else I should have heard of them too.*

There is a great deal of Truth and good Sense in that common Saying and Doctrine of the *Stoics*, though they might carry it too far, that *it is not Things but Thoughts that disturb and hurt us* (g). Now as Self-acquaintance teaches a Man the right Government of the Thoughts, (as is shewn above, Part I. Chap. XIV.) it will help him to expel all anxious, tormenting, and fruitless Thoughts, and retain the most quieting and useful ones; and so keep all easy within. Let a Man but try the Experiment, and he will find, that a little Resolution will make the greatest Part of the Difficulty vanish.

(2.) Self-

(g) Ταρασσει της ανθρωπες, κ τα πραγματα, αλλα τα περι των πραγμαων δογματα. *Id. cap.* 10. *It is not Things, but Mens Opinions of Things that disturb them.*

Μεμνησο ὁτι ουκ ὁ λοιδορων η τυπλων ὑβριζει, αλλα το δογμα το περι τουλων ὡς ὑβριζονλων. *Id. cap.* 27.——*Remember, it is not he that reviles or assaults you, that injures you, but your thinking that they have injured you.*——Σε γαρ αλλος ου βλαψει, αν μη συ θελης· τοτε δε ιση βαβλαμμενος, ὁταν υπολαβης βλαπτεσθαι. *Id. pag.* 37——*No Man can hurt you, unless you please to let him; then only are you hurt when you think yourself so.*

Τα πραγματα ουκ απτεται της ψυχης, αλλ' εξω ἑστηκεν ἁιρεμουντα· αι δε οχλησεις εκ μονης της ενδον υπολη ψεως. *Marc. Anton. Med. lib.* 4. § 3. *Things do not touch the Mind, but stand quietly without; the Vexation comes from within, from our Suspicions only.*——Again, Τα πραγματα αυτα ουδ' ὁπωσουν ψυχης απτεται· ουδε εχει εισοδον προς ψυχην· ουδε τρεψαι ουδε κινησαι ψυχην δυναται· τρεπει δε και κινει αυτη εαυτην μονη. *Id. lib.* 5. § 19. *Things themselves cannot affect the Mind; for they have no Entrance into it, to turn and move it. It is the Mind alone that turns and moves itself.*

Chap. II. *of Self-Knowledge.* 141

(2.) Self-Knowledge will be a good Ballast to the Mind under any accidental *Hurry* or *Disorder* of the *Passions*. It curbs their Impetuosity; puts the Reins into the Hands of Reason; quells the rising Storm, ere it make Shipwreck of the Conscience; and teaches a Man to *leave off Contention before it be meddled with* *, it being much safer to keep the Lion chained, than to encounter it in its full Strength and Fury. And thus will a wise Man, for his own Peace, deal with the Passions of others, as well as his own.

Self-Knowledge, as it acquaints a Man with his Weaknesses and worst Qualities, will be his Guard against them; and a happy Counterballance to the Faults and Excesses of his natural Temper.

(3.) It will keep the Mind sedate and calm under the Surprize of *bad News*, or *afflicting Providences*.

' For am I not a Creature of GOD? And my
' Life and Comforts, are they not wholly at his
' Dispose, from whom I have received them; and
' by whose Favour I have so long enjoyed them;
' and by whose Mercy and Goodness I have still
' so many left?

' A Heathen can teach me, under such Losses
' of Friends, or Estate, or any Comfort, to
' direct my Eyes to the Hand of GOD, by
' whom it was lent me, and is now recalled;
' that I ought not to say, *it is lost*, but *restored*.
' And

---

* *Prov.* xvii. 14.

' And tho' I be injuriously deprived of it, still
' the Hand of GOD is to be acknowledged;
' for what is it to me, by what Means, he that
' gave me that Blessing, takes it from me a-
' gain' (*b*)?

He that rightly *knows himself* will live every Day dependent on the Divine Author of his Mercies, for the Continuance and Enjoyment of them. And will learn from a higher Authority than that of a Heathen Moralist, that he hath nothing he can properly call *his own*, or ought to depend upon as such. That he is but a *Steward* employed to dispense the good Things he possesses, according to the Direction of his Lord, at whose Pleasure he holds them; and to whom he should be ready at any Time chearfully to resign them, *Luke* xvi. 1.

(4.) Self-Knowledge will help a Man to preserve an Equanimity and Self-possession under all the various Scenes of *Adversity* and *Prosperity*.

Both have their Temptations: To some the Temptations of Prosperity are the greatest; to others, those of Adversity. *Self-Knowledge* shews a Man which of these are greatest to *him:* And, at the Apprehension of them, teaches him to arm himself accordingly; that nothing may deprive him of his Constancy and Self-possession, or lead him to act unbecoming the Man or the Christian.

We

(*b*) *Epictet. Enchirid. cap.* 15.

Chap. I.   *of Self-Knowledge.*   143

We commonly say, *no one knows what he can bear, till he is tried.* And many Persons verify the Observation, by bearing Evils much better than they feared. Nay, the Apprehension of an approaching Evil often gives a Man a greater Pain than the Evil itself. This is owing to Inexperience and Self-ignorance,

A Man that knows himself, his own Strength and Weakness, is not so subject as others, to the melancholy Presages of the Imagination; and whenever they intrude, he makes no other use of them than to take the Warning, collect himself, and prepare for the coming Evil; leaving the Degree, Duration, and the Issue of it with Him, who is the sovereign Disposer of all Events, in a quiet Dependence on his Power, Wisdom and Goodness.

Such Self-possession is one great Effect and Advantage of *Self-Knowledge.*

CHAP. II.

Self-Knowledge leads to a wise and steady Conduct.

II. AS *Self-Knowledge will keep a Man calm and equal in his* Temper, *so it will make him wise and cautious in his* Conduct.

A precipitant and rash Conduct is ever the Effect of a confused and irregular Hurry of Thought.

Thought. So that when by the Influence of Self-Knowledge, the Thoughts become cool, sedate and rational, the Conduct will be so too. It will give a Man that even, steady, uniform Behaviour in the Management of his Affairs, that is so necessary for the Dispatch of Business; and prevent many Disappointments and Troubles which arise from the unsuccessful Execution of immature or ill-judged Projects.

In short, most of the Troubles which Men meet with in the World may be traced up to this Source, and resolved into *Self-ignorance*. We may complain of Providence, and complain of Men; but the Fault, if we examine it, will commonly be found to be our own. Our Imprudence, which arises from Self-ignorance, either brings our Troubles upon us, or increases them. Want of Temper and Conduct will make any Affliction double.

What a long Train of Difficulties do sometimes proceed from *one* wrong Step in our Conduct, into which Self-ignorance or Inconsideration betrayed us? And every Evil that befals us in consequence of *that*, we are to charge upon ourselves.

CHAP. III.

Humility the Effect of Self-Knowledge.

III. TRUE *Self-Knowledge always produces Humility.*

Pride is ever the Off-spring of Self-ignorance. The Reason Men are vain and Self-sufficient is, because they do not know their own Failings; and the Reason they are not better acquainted with them is, because they hate Self-inspection. Let a Man but turn his Eyes within, scrutinize himself, and study his own Heart, and he will soon see enough to make him humble. *Behold, I am vile* *, is the Language only of Self-Knowledge (*i*).

Whence is it that young People are generally so vain, Self-sufficient and assured; but because they have taken no Time or Pains to cultivate a Self-acquaintance? And why does Pride and Stiffness appear so often in advanced Age, but because Men grow old in Self-ignorance? A moderate Degree of Self-Knowledge would cure an inordinate Degree of Self-complacency (*k*).

---

* *Job* xi. 4.

(*i*) Qui bene seipsum cognoscit sibi ipsi vilescit, nec laudibus delectatur humanis. *Tho. à Kemp. de Imit. Chr. lib.* **1.** *cap.* 2.

(*k*) Quanto quis minus se videt, tanto minus se displicet. *Greg.*

Humility is not more necessary to Salvation, than Self-Knowledge is to Humility (*l*).

It would effectually prevent that bad Disposition which is too apt to steal upon and infect some of the best human Minds (especially those who aim at singular and exalted Degrees of Piety) *viz.* a religious Vanity or *spiritual Pride.* Which without a good deal of Self-Knowledge and Self-Attention will gradually insinuate into the Heart, taint the Mind, and sophisticate our Virtues before we are aware; and in Proportion to its Prevalence make the Christian Temper degenerate into the *Pharisaical.*

' Might I be allowed to chuse my own Lot,
' I should think it much more eligible to want
' my spiritual Comforts, than to abound in
' these at the Expence of my Humility. No;
' let a penitent and a contrite Spirit be always
' my Portion; and may I ever *so* be the Favou-
' rite of Heaven, as never to forget that I am
' *Chief of Sinners.* Knowledge in the sublime
' and glorious Mysteries of the Christian Faith,
' and ravishing Contemplations of GOD and a
' future State, are most desirable Advantages;
' but still I prefer *Charity which edifieth* before
' the

(*l*) Scio neminem absque sui cognitione salvari, de quâ nimirum mater salutis, humilitas oritur, et timor Domini. *Bernard.* ——— Utraque cognitio Dei, scilicet et tui, tibi necessaria est ad salutem; quia sicut ex notitiâ tui venit in te timor Dei, atque ex Dei notitiâ itidem amor; sic è contra, ex ignorantia tui, superbia, ac de Dei ignorantia venit desperatio. *Idem in Cantic.*

'the highest intellectual Perfections of that *Know-*
'*ledge which puffeth* up *.—Those spiritual Ad-
'vantages are certainly best for us, which increase
'our Modesty and awaken our Caution, and dis-
'pose us to suspect and deny ourselves.———The
'highest in GOD's Esteem, are meanest in their
'own. And their Excellency consists in the
'Meekness and Truth, not in the Pomp and Os-
'tentation of Piety, which affects to be seen and
'admired of Men (*m*).

CHAP.

* 1 *Cor.* viii. 1.
(*m*) *Stanhope's Tho. à Kemp.* B. 2. *cb.* 11.

[CHRIST.] ' My Son, when thou feelest thy Soul warmed
' with Devotion and holy Zeal for my Service, it will be adviseable
' to decline all those Methods of publishing it to the World, which
' vain Men are so industrious to take, and content thyself with its
' being known to GOD and thine own Conscience. Rather endea-
' vour to moderate and suppress those pompous Expressions of it, in
' which some place the very Perfection of Zeal. Think meanly
' of thy own Virtues.——— Some Men of a bold ungoverned Zeal
' aspire at Things beyond their Strength, and express more Vehe-
' mence than Conduct in their Actions. They are perfectly carried
' out of themselves with Eagerness; forget that they are still poor
' Insects upon Earth, and think of nothing less than building their
' Nest in Heaven. Now these are often left to themselves, and
' taught by sad Experience, that the faint Flutterings of Men are
' weak and ineffectual; and that none soars to Heaven except I
' assist his Flight, and mount him on my own Wings.—Virtue
' does not consist in Abundance of Illumination and Knowledge;
' but in Lowliness of Mind, in Meekness, and Charity; in a
' Mind intirely resigned to GOD, and sincerely disposed to serve
' and please him; in a just Sense of every Man's Vileness;
' and not only thinking very meanly of one's self, but be-
'ing

CHAP. IV.

Charity another Effect of Self-Knowledge.

IV. SELF-KNOWLEDGE *greatly promotes a Spirit of* Meekness *and* Charity.

The more a Man is acquainted with his own Failings, the more is he disposed to make Allowances for those of others. The Knowledge he hath of himself, will incline him to be as severe in his Animadversions on his own Conduct, as

"ing well content to be so thought of by others. *Idem,* Book 3. Chap. 8."

" It is a dangerous Drunkenness, I confess, that of Wine; but there is another more dangerous. How many Souls do I see in the World drunk with Vanity, and a high Opinion of themselves? This Drunkenness causes them to make a thousand false Steps, and a thousand Stumbles. Their Ways are all oblique and crooked. Like men in Drink, they have always a great Opinion of their own Wisdom, their Power, and their Prudence; all which often fail them.—— Examine well thyself, my Soul; see if thou art not tainted with this Evil. Alas! if thou deniest it, thou provest it. It is great Pride, to think one has no Pride; for it is to think you are as good indeed, as you esteem yourself. But there is no Man in the World but esteems himself better than he truly is.

" Thou wilt say, it may be, thou hast a very ill Opinion of thyself. But be assured, my Soul, thou dost not despise thyself so much as thou art truly despicable. If thou dost despise thyself indeed, thou makest a Merit of that very Thing; so that Pride is attached to this very Contempt of thyself." *Jurieu's Method of Devot.* pag. 8. ch. 10.

as he is on that of others; and as candid to their Faults as he is to his own *.

There is an uncommon Beauty, Force, and Propriety in that Caution which our Saviour gives us, *And why beholdest thou the Mote that is in thy Brother's Eye, but considerest not the Beam that is in thine own Eye? Or how wilt thou say to thy Brother, let me pull out the Mote out of thine Eye, and behold a Beam is in thine own Eye? Thou Hypocrite, first cast the Beam out of thine own Eye, and then shalt thou see clearly to cast out the Mote out of thy Brother's Eye* †. In which Words these four Things are plainly intimated; (1.) That some are much more quick-sighted to discern the Faults and Blemishes of others, than their own: Can spy a *Mote* in another's Eye, sooner than a *Beam* in their own. (2.) That they are often the most forward to correct and cure the Foibles of others, who are most unqualified for that Office. The Beam in their own Eye makes them altogether unfit to pull out the Mote from their Brother's. A Man half blind himself should never set up for an *Oculist*. (3.) That they who are inclined to deal in Censure should always *begin at Home*. (4.) Great Censoriousness is great *Hypocrisy*. *Thou Hypocrite*, &c. all this is nothing but the Effect of woful Self-ignorance.

* "The great GOD seems to have given that Commandment (KNOW THYSELF) to those Men more especially, who are apt to make Remarks on other Men's Actions, and forget themselves." *Plutarch's Mor. Vol.* i, *pag.* 273.

† *Mat.* vii. 3.—5.

This common Failing of the Human Nature the Heathens were very sensible of (*n*); and imag'd it in the following Manner. Every Man (say they) carries a Wallet, or two Bags with him; the one hanging before him, and the other behind him; into that *before*, he puts the Faults of others; into that *behind*, his own; by which Means he never sees his own Failings, whilst he has those of others always before his Eyes (*o*).

But Self-Knowledge now helps us to turn this *Wallet*; and place that which hath our own Faults, before our Eyes, and that which hath in it those of others, behind our Back. A very necessary Regulation this, if we would behold our own Faults in the same Light in which they do. For we must not expect that others will be as blind to our Foibles as we ourselves are. They will carry them before their Eyes, whether we do or no. And to imagine that the World takes no Notice of them, because we do not, is just as wise as to fancy that others do not see us, because we shut our Eyes.

CHAP.

(*n*) ——Egomet mî ignosco Mænius inquit;
Stultus et improbus hic amor est, dignusque notari.
Cum tua pervideas oculis male Lippus inunctis,
Cur in amicorum vitiis tam cernis acutum
Quâm aut aquila, aut Serpens *Epidaurius?*
Hor. Sat. 3. lib. 1.

Fit enim, nescio quomodo, ut magis in aliis cernamus quam in nobismet ipsis, siquid delinquitur. *Cicero.*

(*o*) Sed præcedenti spectatur mantica tergo. *Per. Sat.* 4.
Non videmus id manticæ quod in tergo est. *Catul. Carm.* 22.

Nostram peram non videntes, aliorum (juxta *Persium*) manticam consideramus. *D. Hieron. Epist.* 91.

CHAP. V.

Moderation the Effect of Self-Knowledge.

V. ANOTHER *genuine Offspring of Self-Knowledge is* Moderation.

This indeed can hardly be conceived to be separate from that of Meekness and Charity beforementioned; but I choose to give it a distinct Mention, because I consider it under a different View and Operation, *viz.* as that which guards and influences our Spirits in all Matters of *Debate* and *Controversy*.

Moderation is a great and important Christian Virtue, very different from that bad Quality of the Mind under which it is often misrepresented and disguised, *viz. Lukewarmness* and Indifference about the Truth. The former is very consistent with a regular and well-corrected Zeal, the latter consists in the *total Want* of it; the former is sensible of, and endeavours with Peace and Prudence to maintain the Dignity and Importance of Divine Doctrines, the latter hath no Manner of Concern about them; the one feels the secret Influences of them, the other is quite a Stranger to their Power and Efficacy; the one laments in secret the sad Decay of vital Religion, the other is an Instance of it. In short, the one proceeds from true Knowledge, the other from great Ignorance; the one is a good Mark of

Sincerity, and the other a certain Sign of Hypocrisy. And to confound two Things together, which are so essentially different, can be the Effect of nothing but great Ignorance, Inconsideration, or an over-heated, injudicious Zeal.

A self-knowing Man can easily distinguish between these two. And the Knowledge which he has of Human Nature in general, from a thorough Contemplation of his own in particular, shews him the Necessity of preserving a Medium (as in every Thing else, so especially) between the two Extremes of a *bigotted Zeal* on the one Hand, and *indolent Lukewarmness* on the other. As he will not look upon every Thing to be worth contending for, so he will look upon nothing worth losing his Temper for in the Contention. Because, though the Truth be of ever so great Importance, nothing can do a greater Disservice to it, or make a Man more incapable of defending it, than intemperate Heat and Passion; whereby he injures and betrays the Cause he is over-anxious to maintain. *The Wrath of Man worketh not the Righteousness of GOD* [*].

Self-Knowledge heals our Animosities, and greatly cools our Debates about Matters of dark and doubtful Speculation. One who knows himself sets too great a Value upon his Time and Temper, to plunge rashly into those vain and fruitless Controversies, in which one of them is sure to be lost, and the other in great Danger of being

---

[*] *James* i. 20.

being so; especially when a Man of bad Temper and bad Principles is the Opponent; who aims rather to *silence* his Adversary with over-bearing Confidence, dark unmeaning Language, authoritative Airs, and hard Words, than *convince* him with solid Argument; and who plainly contends not for Truth but Victory. Little Good can be done to the best Cause in such a Circumstance. And a wise and moderate Man, who knows Human Nature, and knows himself, will rather give his Antagonist the Pleasure of an imaginary Triumph, than engage in so unequal a Combat.

An Eagerness and Zeal for Dispute, on every Subject, and with every one, shews great Self-sufficiency; that never-failing Sign of great Self-ignorance. And true Moderation, which creates an Indifference to little Things, and a wise and well-proportioned Zeal for Things of Importance, can proceed from nothing but true Knowledge, which has its Foundation in Self-acquaintance.

CHAP. VI.

Self-Knowledge improves the Judgment.

VI. ANOTHER great *Advantage of being well acquainted with ourselves is, that it helps us to form a better Judgment of other Things.*

Self-Knowledge indeed does not inlarge or increase our natural Capacities, but it guides and regulates them; leads us to the right Use and Application of them; and removes a great many Things which obstruct their due Exercise, as Pride, Prejudice, and Passion, &c. which oftentimes so miserably pervert the rational Powers.

He that hath taken a just Measure of himself, is thereby better able to judge of other Things.

(1.) He knows how to judge of *Men* and *Human Nature* better. ——— For Human Nature, setting aside the Difference of natural Genius, and the Improvements of Education and Religion, is pretty much the same in all. There are the same Passions and Appetites, the same natural Infirmities and Inclinations in all Mankind; tho' some are more predominant and distinguishable in some, than they are in others. So that if a Man be but well acquainted with his own, *this*, together with a very little Observation on Human Life, will soon discover to him those of other Men; and shew him very impartially their particular

Fail-

Chap. VI. *of Self-Knowledge.*

Failings and Excellencies, and help him to form a much truer Sentiment of them, than if he were to judge only by their *Exterior*, the Appearance they make in the Eye of the World, or the Character given of them by others; both which are often very fallacious.

(2.) Self-Knowledge will teach us to judge rightly of *Facts* as well as Men. It will exhibit Things to the Mind in a proper Light, and true Colours, without those false Glosses and Appearances which Fancy throws upon them, or in which the Imagination often paints them. It will teach us to judge not with the Imagination, but with the Understanding; and will set a Guard upon the former, which so often represents Things in wrong Views, and gives the Mind false Impressions. See *Part* I. *Chap.* IV.

(3.) It helps us to estimate the true Value of all *worldly good Things*. It rectifies our Notions of them, and lessens that enormous Esteem we are apt to have for them. For when a Man knows himself, and his true Interests, he will see how far, and in what Degree, these Things are suitable to him, and subservient to his Good; and how far they are unsuitable, insnaring, and pernicious. *This*, and not the common Opinion of the World, will be his Rule of Judgment concerning them. By this he will see quite through them; see what they really are at Bottom; and how far a wise Man ought to desire them. The Reason why Men value them so extravagantly is, because

because they take but a superficial View of them, and only look upon their Out-side, where they are most showy and inviting. Were they to look within them, consider their intrinsic Worth, their ordinary Effects, their Tendency and their End, they would not be so apt to over-value them. And a Man that has learned to see through himself, can easily see through these (*p*).

CHAP. VII.

Self-Knowledge directs to the proper Exercise of Self-denial.

VII. A MAN *that knows himself best, knows how, and wherein, he ought to deny himself.*

The great Duty of *Self-denial*, which our Saviour so expressly requires of all his Followers, (plain and necessary as it is) has been much mistaken

---

(*p*) Abstrahunt a recto divitiæ, honores, potentia, et cætera quæ opinione nostrâ chara sunt, precio suo vilia. Nescimus æstimare res: de quibus, non cum famâ sed cum rerum naturâ, deliberandum est. Nihil habent ista magnificum, quo mentes in se nostras trahant, præter hoc quod mirari illa consuevimus. Non, enim quia concupiscenda sunt, laudantur, sed concupiscuntur quia laudata sunt. Sen. Epist. 82.————*Riches, Honours, Power, and the like, which owe all their Worth to our false Opinion of them, are too apt to draw the Heart from Virtue. We know not how to prize them; they are not to be judged of by the common Vogue, but by their own Nature. They have nothing to attract our Esteem, but that we are used to admire them; they are not cryed up because they are Things that ought to be desired, but they are desired because they are generally cryed up.*

taken and abused; and that not only by the Church of *Rome*, in their Doctrines of Penance, Fasts, and Pilgrimages, but by some Protestant Christians in the Instances of voluntary Abstinence, and unnecessary Austerities. Whence they are sometimes apt to be too censorious against those who indulge themselves in the Use of those indifferent Things, which *they* make it a Point of Conscience to abstain from. Whereas would they confine their Exercise of Self-denial to the plain and important Points of Christian Practice, devoutly performing the necessary Duties they are most averse to, and resolutely avoiding the known Sins they are most inclined to, under the Direction of Scripture, they would soon become more solid, judicious, and exemplary Christians; and did they know themselves, they would easily see that herein there is Occasion and Scope enough for Self-denial; and that to a Degree of greater Severity and Difficulty than there is in those little corporal Abstinences and Mortifications they enjoin themselves.

(1.) Self-Knowledge will direct us to the necessary Exercises of Self-denial, with regard to the *Duties* our Tempers are most averse to.

There is no one, but, at some Times, finds a great Backwardness and Indisposition to some Duties which he knows to be seasonable and necessary. This then is a proper Occasion for Self-discipline. For to indulge this Indisposition is very dangerous, and leads to an habitual Neglect

of

of known Duty; and to refist and oppofe it, and to prepare for a diligent and faithful Difcharge of the Duty, notwithftanding the many Pleas and Excufes that carnal Difpofition may urge for the Neglect of it, this requires no fmall Pains and Self-denial; and yet it is very neceffary to the Peace of Confcience.

As for our Encouragement to this Piece of Self-denial, we need only remember that the Difficulty of the Duty, and our Unfitnefs for it, will, upon the Trial, be found to be much lefs than we apprehended. And the Pleafure of reflecting, that we have difcharged our Confciences, and given a frefh Teftimony of our Uprightnefs, will more than compenfate the Pains and Difficulty we found therein. And the oftener thefe criminal Propenfions to the wilful Neglect of Duty are oppofed and conquered, the feldomer will they return, or the weaker will they grow. Till at laft, by Divine Grace, they will be wholly overcome; and in the Room of them will fucceed an habitual *Readinefs to every good Work* *, and a very fenfible Delight therein: A much happier Effect than can be expected from the feverelt Exercifes of Self-denial, in the Inftances before-mentioned.

(2.) A Man that knows himfelf will fee an equal Neceffity for Self-denial, in order to check and controul his *Inclinations to finful Actions:* To fubdue the Rebel within; to refift the Solicitations

---

* *Tit.* iii. 1.

Chap. VII. *of Self-Knowledge.*

ons of Sense and Appetite; to summon all his Wisdom to avoid the Occasions and Temptations to Sin, and all his Strength to oppose it.

All this (especially if it be a favourite constitutional Iniquity) will cost a Man Pains and Mortification enough. For Instance, the subduing a violent Passion, or taming a sensual Inclination, or forgiving an apparent Injury and Affront. It is evident, such a Self-conquest can never be attained without much Self-Knowledge, and Self-denial.

And that Self-denial that is exercised this Way, as it will be a better Evidence of our Sincerity, so it will be more helpful and ornamental to the Interests of Religion, than the greatest Zeal in those particular Duties which are most suitable to our natural Tempers, or than the greatest Austerities in some particular Instances of Mortification, which are not so necessary, and perhaps not so difficult or disagreeable to us as this.

To what amazing Heights of Piety may some be thought to mount, (raised on the Wings of a flaming Zeal, and distinguished by uncommon Precisenefs and Severity about little Things) who all the while, perhaps, cannot govern *one Passion*, and appear yet ignorant of, and Slaves to, their darling Iniquity! Through an ignorance of themselves, they misapply their Zeal, and misplace their Self-denial; and by that Means blemish their Characters with a visible Inconsistency (*q*.)

CHAP.

(*q*) *A pious Zeal may be active and yet not pernicious, and shine without burning. Intemperate Zeal is like* Sirius *in* Homer.

Δαμψίςατος

CHAP. VIII.

Self-Knowledge promotes our Usefulness in the World.

VIII. THE *more we know of ourselves, the more* useful *we are like to be, in those Stations of Life in which Providence hath fixed us.*

When we know our proper Talents and Capacities, we know in what Manner we are capable of being useful; and the Consideration of our Characters and Relations in Life, will direct us to the proper Application of those Talents; show us to what Ends they were given us, and to what Purposes they ought to be improved.

'Many of those who set up for Wits, and pre-
' tend to a more than ordinary Sagacity and De-
' licacy of Sense, do notwithstanding spend their
' Time *unaccountably*; and live away whole Days,
' Weeks, and sometimes Months together, to as
' little Purpose, (tho' it may be not so innocent-
' ly) as if they had been asleep all the while.——
                                    ' But

Λαμπρότατος μὲν ὅγ' ἐςὶ, κακὸν δέ τε σῆμα τέτυκται
Καί τε φέρει πολλὸν πυρετὸν δειλοῖσι βροτοῖσι·
Ille quidem clarâ, sed sævâ luce coruscat,
Et morbos æstusque adfert mortalibus ægris.   *Il.* x. 30.

  Pious Zeal is like the gentle Flame *in* Virgil.
Ecce levis subito de vertice visus Jüli
Fundere lumem apex, tractuque innoxia mollis
Lambere flamma comas, et circum tempora pasci.
                                    Æn. II.
                          Jortin's Disc. p. 31.

'But if their Parts be so good as they would have
'others believe, sure they are *worth* improving;
'if not, they have the more Need of it.————
'Greatness of Parts is so far from being a Dis-
'charge from Industry, that I find Men of the
'most exquisite Sense, in all Ages, were always
'most curious of their Time. And therefore I
'very much suspect the Excellency of those Mens
'Parts, who are dissolute and careless Misspenders
'of it' (r).

It is a sad Thing to observe, how miserably some Men debase and prostitute their Capacities. Those Gifts and Indulgencies of Nature, by which they outshine many others, and by which they are capable of doing real Service to the Cause of Virtue and Religion, and of being eminently useful to Mankind, they either entirely neglect, or shamefully abuse, to the Dishonour of GOD, and the Prejudice of their Fellow-Creatures, by incouraging and imboldening them in the Ways of Vice and Vanity. For the false Glare of a profane Wit will sometimes make such strong Impressions on a weak, unsettled Mind, as to overbear the Principles of Reason and Wisdom, and give it too favourable Sentiments of what it before abhorred. Whereas the same Force and Sprightliness of Genius would have been very happily and usefully employed in putting Sin out of Countenance, and in rallying the Follies, and exposing the Inconsistencies of a vicious and profligate Character.

The

(r) *Norris's Misc.* p. 120.

The more Talents and Abilities Men are blessed with, the more Pains they ought to take.—This is *Chrysostom*'s Observation. And the Reason is obvious; because they have more to answer for, than other Men, which I take to be a better Reason than what is assigned by this Father, *viz. because they have more to lose (s).*

When a Man once knows where his Strength lies, wherein he excels, or is capable of excelling, how far his Influence extends, and in what Station of Life Providence hath fixed him, and the Duties of that Station; he then knows what Talents he ought to cultivate, in what Manner and to what Objects they are to be chiefly directed and applied, in order to shine in that Station, and be useful in it. This will keep him even and steady in his Pursuits and Views; consistent with himself, uniform in his Conduct, and useful to Mankind; and will prevent his shooting at a wrong Mark, or missing the right one he aims at; as Thousands do, for want of this necessary Branch of Self-Knowledge. See *Part* I. *Chap.* V.

CHAP.

(s) Ὥστε τοῖς σοφωτέροις, μᾶλλον ἢ τοῖς ἀμαθεστέροις, μείζων ὁ πόνος· οὐδὲ γὰρ ὑπὲρ τῶν αὐτῶν ἡ ζημία, ἀμελοῦσι τούτοις κἀκείνοις. *De Sacred. l.* v. c. 5.

CHAP. IX.

Self-Knowledge leads to a Decorum and Consistency of Character.

IX. A MAN *that knows himself, knows how to act with* Discretion *and* Dignity *in every Station and Character.*

Almost all the *Ridicule* we see in the World takes its Rise from Self-ignorance. And to this Mankind by common Assent ascribe it; when they say of a Person that acts out of Character, *he does not know himself.* Affectation is the Spring of all Ridicule, and Self-ignorance the true Source of Affectation. A Man that does not know his proper Character, nor what becomes it, cannot act suitably to it. He will often affect a Character that does not belong to him; and will either act above or beneath himself, which will make him equally contemptible in the Eyes of them that know him (*t*).

A Man of *superior* Rank and Character, that knows himself, knows that he is but a Man; subject to the same Sicknesses, Frailties, Disappointments, Pains, Passions, and Sorrows, as other Men; that true Honour lies in those Things, in which it is possible for the meanest Peasant to excel him; and therefore he will not be vainly arrogant

---

(*t*) Omnique in re posse quod deceat facere, artis et naturæ est; scire, quid, quandoque deceat, prudentiæ. *Cic. de Orat. l.* 3. § 55.

rogant. He knows that they are only transitory and accidental Things, that set him above the rest of Mankind; that he will soon be upon a Level with them; and therefore learns to condescend: And there is a Dignity in this Condescension; it does not sink, but exalt, his Reputation and Character.

A Man of *inferior* Rank, that knows himself, knows how to be content, quiet, and thankful, in his lower Sphere. As he has not an extravagant Veneration and Esteem for those external Things which raise one Man's Circumstances so much above another's, so he does not look upon himself as the worse or less valuable Man, purely because he has them not; much less does he envy them that have them. As he has not their Advantages, so neither has he their Temptations: He is in that State of Life, which the great Arbiter and Disposer of all Things hath allotted him; and he is satisfied: But as a Deference is owing to external Superiority, he knows how to pay a proper Respect to those that are above him, without that abject and servile Cringing, which discovers an inordinate Esteem for their Condition. As he does not over-esteem them for those little accidental Advantages in which they excel him, so neither does he overvalue himself for those Things in which he excels others.

Were *Hearers* to know themselves, they would not take upon them to dictate to their Preachers; or teach their Ministers how to teach *them*;
(which,

(which, as St. *Austin* observes, (*u*) is the same Thing as if a Patient, when he sends for a Physician, should prescribe to him what he would have him prescribe;) but, if they happen to hear something not quite agreeable to their former Sentiments, would betake themselves more diligently to the Study of their Bibles, to know *whether those Things were so*.*

And were *Ministers* to know themselves, they would know the Nature and Duty of their Office, and the Wants and Infirmities of their Hearers better, than to domineer over their Faith, or shoot over their Heads, and seek their own Popularity, rather than *their* Benefit. They would be more solicitous for their Edification, than their Approbation; (the most palatable Food is not always the most wholesome;) and like a faithful Physician, would earnestly intend and endeavour their Good, though it be in a Way they may not like; and rather risk their own Characters with weak and captious Men, than *with-hold any Thing that is needful for them*, or be unfaithful to GOD and their own Consciences. Patients must not expect to be always *pleased*, nor Physicians to be always *applauded*.

CHAP.

(*u*) Norit medicus quid salutiferum, quidve contrarium petat ægrotos. Ægroti estis, nolite ergo dictare quæ vobis medicamina velit opponere.

* *Acts* xvii. 11.

CHAP. X.

Piety the Effect of Self-Knowledge.

X. SELF-KNOWLEDGE *tends greatly to cultivate a spirit of true* Piety.

Ignorance is so far from being the *Mother of Devotion*, that nothing is more destructive of it. And of all Ignorance, none is a greater Bane to it than *Self-ignorance*. This indeed is very consistent with Superstition, Bigotry, and Enthusiasm, those common *Counterfeits* of Piety, which by weak and credulous Minds are often mistaken for it. But true Piety and real Devotion can only spring from a just Knowledge of GOD and ourselves; and the Relation we stand in to him, and the Dependence we have upon him. For when we consider ourselves as the Creatures of GOD, whom he made for his Honour, and as Creatures incapable of any Happiness, but what results from his Favour; and as intirely and continually dependent upon him for every Thing we have and hope for; and whilst we bear this Thought in our Minds, what can induce or prompt us more to love and fear and trust Him, as our GOD, our Father, and all-sufficient Friend and Helper?

CHAP. XI.

Self-Knowledge teaches us rightly to perform the Duties of Religion.

XI. SELF-KNOWLEDGE *will be a good Help and Direction to us in many of our* Devout *and* Christian *Exercises.* Particularly,

(1.) In the Duty of *Prayer*; both as to the *Matter* and *Mode* (*w*).———He that rightly knows himself, will be very sensible of his spiritual Wants; and he that is well acquainted with his spiritual Wants, will not be at a Loss what to pray for. " Our Hearts would be the best Pray-
" er-Books, if we were well skilful in reading
" them. Why do Men pray, and call for Pray-
" ers when they come to die, but that they be-
" gin a little better to know themselves? And
" were they now but to hear the Voice of GOD
" and Conscience, they would not remain speech-
" less. But they that are born deaf are always
" dumb"(*x*).

Again, Self-Knowledge will teach us to pray, not only with Fluency, but Fervency; will help us to *keep the Heart*, as well as *order our Speech* before GOD; and so promote the Grace as well

---

(*w*) Ille Deo veram orationem exhibet qui semitipsum cognoscit. *Greg.*

well as Gift of Prayer. Did we but seriously consider what we are, and what we are about: whom we pray to, and what we pray for, it is impossible we should be so dead, spiritless, and formal in this Duty, as we too often are. The very Thought would inspire us with Life, and Faith, and Fervour.

(2) Self-Knowledge will be very helpful to us in the Duty of *Thanksgiving*: As it shews us both how suitable and how seasonable the Mercies are which we receive. A Christian that keeps up an Intelligence with himself, considers what he hath, as well as what he wants; and is no less sensible of the Value of his Mercies, than his Unworthiness of them: And this is what makes him thankful. For this Reason it is, that one Christian's Heart even melts with Gratitude for those very Mercies, which others disesteem and depreciate; and perhaps despise, because they have not what they think greater. But a Man that knows himself, knows that he deserves nothing, and therefore is thankful for every Thing. For Thankfulness as necessarily flows from Humility, as Humility does from Self-acquaintance.

(3.) In the Duties of *reading and hearing* the Word of GOD. Self-Knowledge is of excellent Use to enable us to *understand* and *apply* that which we read or hear. Did we understand our Hearts better, we should *understand* the Word of GOD better; for that speaks to the Heart. A Man that is acquainted with his own Heart, prefently fees how deeply the Divine Word penetrates and explores, fearches and lays open its moſt inward Parts. He feels what he reads; and finds that *a quickening Spirit*, which to a felf-ignorant Man is but a *dead Letter*.

Moreover, this Self-acquaintance teaches a Man to *apply* what he reads and hears of the Word of GOD. He fees the Pertinence, Congruity, and Suitableneſs of it to his own Cafe; and lays it up faithfully in the Store-Room of his Mind, to be digeſted and improved by his Afterthoughts. And it is by this Art of applying Scripture, and urging the moſt fuitable Inſtructions and Admonitions of it home upon our Confciences, that we receive the greateſt Benefit by it.

(4.) Nothing is of more eminent Service in the great Duty of *Meditation*; eſpecially in that Part of it which confiſts in Heart converfe. A Man, who is unacquainted with himſelf, is as unfit to converfe with his Heart, as he is with a Stranger he never faw, and whoſe Taſte and Temper he is altogether unacquainted with. He knows not how to get his Thoughts about him. And when he has, he knows not how to range and fix them; and hath no more the Command of them, than a General has of a wild undifciplined Army, that has never been exerciſed, or accuſtomed to Obedience and Order. But one, who hath made it the Study of his Life to be acquainted with

himſelf,

himself, is soon disposed to enter into a free and familiar Converse with his own Heart; and in such a Self-conference improves more in true Wisdom, and acquires more useful and substantial Knowledge, than he could do from the most polite and refined Conversation in the World. —— Of such excellent Use is Self-Knowledge in all the Duties of Devotion and Piety.

CHAP. XII.

Self-Knowledge the best Preparation for Death.

XII. SELF-KNOWLEDGE *will be an habitual* Preparation for Death, *and a constant Guard against the Surprise of it.* Because it fixes and settles our Hopes of future Happiness. That which makes the Thoughts of Death so terrifying to the Soul, is its utter Uncertainty what will become of it after Death. Were this Uncertainty to be removed, a thousand Things would reconcile us to the Thoughts of dying (*y*).

"Distrust

(*y*) Illa quoque res morti nos alienat, quod hæc jam novimus, illa ad quæ transituri sumus, nescimus qualia sint. Et horremus ignota. Naturalis præterea tenebrarum metus est, in quas adductura mors creditur. Sen. Epist. 83. *It is this makes us averse to Death, that it translates us to Objects we are unacquainted with, and we tremble at the Thoughts of those Things that are unknown to us. We are naturally afraid of being in the Dark; and Death is a Leap in the Dark.*

Chap. XII.   *of Self-Knowledge.*

"Distrust and Darkness of a future State,
"Is that which makes Mankind to dread their
                                                                 (Fate;
"Dying is nothing; but 'tis this we fear,
"To be we know not *what*, we know not
                                                                (*where.*"

Now Self-Knowledge, in a good Degree, dissipates this Gloom, and removes this dreadful Doubt. For as the Word of God hath revealed the Certainty of a future State of Happiness, which the good Man shall enter upon after Death, and plainly described the requisite Qualifications for it; when by a long and laborious Self-acquaintance, he comes distinctly to discern those Qualifications in himself, his Hopes of Heaven soon raise him above the Fears of Death. And tho' he may not be able to form any clear or distinct Conception of the Nature of that Happiness, yet in general he is assured that it will be a most exquisite and extensive one, and will contain in it every Thing necessary to make it complete; because it will come immediately from GOD himself(z). Whereas they who know not what they

are,

(z) ' When we say that the State of the other World is unknown, the only Meaning of it is, that it is a State of such Happiness, so far beyond any Thing we ever yet experienced, that we cannot form any Notion or Idea of it: We know that there is such a Happiness; we know in some Measure wherein this Happiness consists; *viz.* in seeing GOD and the blessed Jesus, who loved us, and gave himself for us; in
' praising

are, must necessarily be ignorant what they *shall be*. A Man that is all Darkness within, can have but a dark Prospect forward (*a*).

O, what would we not give for *solid* Hope in Death! Reader, wouldst thou have it, *know GOD*, and *know thyself*.

'praising our Creator and Redeemer; in conversing with Saints and Angels. But how great, how ravishing and transporting a Pleasure this is, we cannot tell, because we never yet felt it.— Now methinks this should not make the Thoughts of Death uneasy to us, should not make us unwilling to go to Heaven; that the Happiness of Heaven is too great for us to know, or to conceive in this World. For Men are naturally fond of unknown and untried Pleasures; which is so far from being a Disparagement to them, that it raises our Expectations of them, that they are unknown. In the Things of this World, Enjoyment usually lessens our Esteem and Value for them, and we always value that most which we have never tried; and methinks the Happiness of the other World should not be the only Thing we despise before we try it.——It is some Encouragement to us that the Happiness of Heaven is too big to be known in this World; for did we perfectly know it now, it could not be very great.'

Sherlock *on Death*, *p*. 71, 72.

(*a*) Illi mors gravis incubat.
　　Qui, notis nimis omnibus,
　　Ignotus moritur sibi.　　　　*Sen. Tba. Thyes.*

Who, expos'd to others Eyes,
Into his own Heart never pries,
Death's to him a strange Surprise.

A TREATISE OF SELF-KNOWLEDGE.

PART III.

Shewing how Self-Knowledge is to be attained.

FROM what hath been said under the two former Parts of the Subject, *Self-Knowledge* appears to be in itself so excellent, and in its Effects so extensively useful and conducive to the Happiness of Human Kind, that nothing need further be added by Way of Motive or Inducement to excite us to make it the great Object of our Study and Pursuit. If we regard our present Peace, Satisfaction, and Usefulness, or our future and everlasting Interests, we shall certainly value and prosecute this Knowledge above all others; as what will be most ornamental to our Characters, and beneficial to our Interest in every State of Life, and abundantly recompence all our Labour.

Were there need of any further Motives to excite us to this, I might lay open the many dreadful Effects of *Self-ignorance*, and shew how plainly it appears to be the original Spring of all the Follies and Incongruities we see in the Characters of Men, and of most of the Mortifications and Miseries they meet with here. This would soon appear by only mentioning the Reverse of those Advantages before specified, which result from *Self-Knowledge*. For what is it, but a Want of Self-Knowledge and Self-government that makes us so unsettled and volatile in our Dispositions? So subject to Transport and Excess of Passions in the varying Scenes of Life? So rash and unguarded in our Conduct? So vain and Self-sufficient? So censorious and malignant? So eager and confident? So little useful in the World, in comparison of what we might be? So inconsistent with ourselves? So mistaken in our Notions of true Religion? So generally indisposed to, or unengaged in the holy Duties of it? And finally, so unfit for Death, and so afraid of dying?—— I say, to what is all this owing, but *Self-ignorance?* The first and fruitful Source of all this long Train of Evils. —— And indeed there is scarce any, but what may be traced up to it. In short, it brutifies Man to be ignorant
of

is to be attained.

of himself. *Man that is in Honour, and understandeth not* (himself especially) *is as the Beasts that perish* †.

"Come home then, O my wandering, self-
"neglecting Soul; lose not thyself in a Wilder-
"ness or Tumult of impertinent, vain, distract-
"ing Things. Thy Work is nearer thee; the
"Country thou shouldst first survey and travel is
"within thee; from which thou must pass to that
"above thee; when by losing thyself in this with-
"out thee, thou wilt find thyself before thou art
"aware in that below thee.——Let the Eyes of
"Fools be in the Corners of the Earth; leave it
"to Men beside themselves, to live as without
"themselves; do thou keep at Home and mind
"thine own Business. Survey thyself, thine own
"Make and Nature, and thou wilt find full Em-
"ploy for all thy most active Thoughts *. But
"dost thou delight in the Mysteries of Nature?
"Consider well the Mystery of thy own. The
"Compendium of all thou studiest is near thee,
"even within thee; thyself being the Epitome
"of the World (*b*).——If either Necessity or
"Duty

---

† *Psal.* xlix. 20.

* Mirantur aliqui altitudines montium, ingentes fluctus maris, altissimos lapsus fluminum, et oceani ambitum, et gyros syderum, et relinquunt seipsos, nec mirantur; saith Saint *Augustin*. *Some Men admire the Heights of Mountains, the huge Waves of the Sea, the steep Falls of Rivers, the Compass of the Ocean, and the Circuit of the Stars, and pass by themselves without Admiration.*

(*b*) Τις ἂν ἀξίως θαυμάσειε την συγγένειαν τοῦ τε ζῶντε συνδέοντος ἐν ἑαυτῷ τὰ θνητὰ τοῖς ἀθανάτοις, καὶ τὰ λογικὰ τοῖς ἀλόγοις συνάπτοντος, τοῦ φέροντος ἐν τῇ καθ' ἑαυτὸν φύσει τῆς πάσης

"Duty, Nature or Grace, Reason or Faith, in-
"ternal Inducements, external Impulses, or eter-
"nal Motives, might determine the Subject of
"thy Study and Contemplation, thou wouldst call
"home thy distracted Thoughts, and employ them
"more on thyself and thy GOD (c)."

Now then let us resolve that henceforth the Study of ourselves shall be the Business of our Lives. That, by the Blessing of GOD, we may arrive at such a Degree of *Self-Knowledge*, as may secure to us the excellent Benefits before-mentioned. To which End we should do well to attend diligently to the Rules laid down in the following Chapters.

CHAP.

της κτισεως την εικονα, δι α και μικρος κοσμος ειρηται, τα τε ταυτης αξιομενα παρα τα Θεα προνοιας; δι ον παντα και τα νυν, και τα μελλοντα· δι ον ο Θεος ανθρωπος γεγονε. *Nem. de Nat. Hom. cap.* 1. *pag.* 34. Who can sufficiently admire the noble Nature of that Creature Man, who hath in him the mortal and the immortal, the rational and irrational Natures united, and so carries about with him the Image of the whole Creation; whence he is called *Microcosm*, or the little World; for whose Sake (so highly is he honoured by GOD) all Things are made both present and future; nay, for whose Sake GOD himself became Man? ——— So that it was not unjustly said by *Gregory Nessene*, that Man was the *Macrocosm*, and the World without the *Microcosm*.

(c) Baxter's *Mischief of Self-ignorance*.

CHAP. I.

Self-examination neceſſary to Self-Knowledge.

I. THE *firſt Thing neceſſary to Self-Knowledge is* Self-inſpection.

We muſt often look into our Hearts if we would know them. They are very deceitful; more ſo than we can imagine till we have ſearched and tried and watched them well. We may meet with Frauds and faithleſs Dealings from Men; but after all, our own Hearts are the greateſt Cheats; and there are none we are in greater Danger from than ourſelves. We muſt firſt ſuſpect ourſelves, then examine ourſelves, then watch ourſelves, if we expect ever to know ourſelves. How is it poſſible there ſhould be any *Self-acquaintance* without *Self-converſe?*

Were a Man to accuſtom himſelf to ſuch Self-employment, he need not live *till Thirty before he ſuſpects himſelf a Fool, or till Forty before he knows it (d)*.

Men could never be ſo bad as they are, if they did but take a proper Care and Scope in this Buſineſs of Self-examination *(e).* If they did but

---

(d) See *the Complaint, or Night-Thoughts,* Part i. *pag.* 28.

(e) Hoc nos peſſimos facit, quod nemo vitam ſuam reſpicit. Quid facturi ſimus, cogitamus, et id raro; quid fecerimus, non cogitamus. *Sen. Epiſt.* 84.

but look backwards to what they were, inwards to what they are, and forwards to what they shall be.

And as this is the first and most necessary Step to Self-acquaintance, it may not be amiss to be a little more particular in it. Therefore,

(1.) This Business of Self-scrutiny must be performed with great *Care* and *Diligence*. Otherwise our Hearts will deceive us, even whilst we are examining them. " When we set ourselves " to think, some Trifle or other presently inter- " rupts and draws us off from any profitable Re- " collection. Nay we ourselves fly out, and are " glad to be diverted from a severe Examination " into our own State; which is sure, if diligently " pursued, to present us with Objects of Shame " and Sorrow, which will wound our Sight, " and soon make us weary of this necessary " Work " (*f*).

Do not let us flatter ourselves then that this is a mighty easy Business. Much Pains and Care are necessary sometimes to keep the Mind intent; and more to keep it impartial. And the Difficulty of it is the Reason that so many are averse to it; and care not to descend into themselves (*g*).

Reader, try the Experiment; retire now into thyself; and see if thou canst not strike out some Light

---

(*f*) Stanhope's *Tho. à Kempis*, pag. 166.
(*g*) Ut nemo in sese tentat descendere! *Pers. Sat.* 4.

Light within, by closely urging such Questions as these——"What am I? For what was I made? "And to what Ends have I been preserved so "long, by the Favour of my Maker? Do I re- "member, or forget those Ends? Have I an- "swered or perverted them?—— What have I "been doing since I came into the World? What "is the World or myself the better for my living "so many Years in it?——What is my allowed "Course of Actions? Am I sure it will bear the "future Test?—Am I now in that State I shall "wish to die in? And, O my Soul, think, and "think again what it is to die. —— Do not put "that most awful Event far from thee; nor pass "it by with a superficial Thought. Canst thou "be too well fortified against the Terrors of "that Day! And art thou sure that the Props, "which support thee now will not fail thee then? "—— What Hopes hast thou for Eternity! "Hast thou indeed that Godly Temper, which "alone can fit thee for the Enjoyment of GOD? "—Which World art thou most concerned for? "What Things do most deeply affect thee?—— "O my Soul, remember thy Dignity; think how "soon the Scene will shift. Why shouldst thou "forget that thou art immortal?"

(2.) This Self-excitation and Scrutiny must be *frequently* made. —— They who have a great deal of important Business on their Hands should often look over their Accounts, and frequently adjust them; lest they should be going backwards, and not know it. And Custom will soon take off the Difficulty of this Duty, and make it delightful.

In our *Morning* Retreat, it will be proper to remember, that we cannot preserve throughout the Day that calm and even Temper we may then be in. That we shall very probably meet with some Things to ruffle us; some Attack on our weak Side. Place a Guard there now. Or however, if no Incidents happen to discompose us, our Tempers will vary; our Thoughts will flow pretty much with our Blood; and the Dispositions of the Mind be a good deal governed by the Motions of the animal Spirits; our Souls will be serene or cloudy, our Tempers volatile or flegmatick, and our Inclinations sober or irregular, according to the Briskness or Sluggishness of the Circulation of the animal Fluids, whatever may be the natural and immediate Cause of that; and therefore we must resolve to avoid all Occasions that may raise any dangerous Ferments there; which, when once raised, will excite in us very different Thoughts and Dispositions from those we now have; which, together with the Force of a fair Opportunity and urgent Temptation, may overset our Reason and Resolution, and betray us into those sinful Indulgences which will wound the Conscience, stain the Soul, and create bitter Remorse in our cooler Reflections. Pious Thoughts

Thoughts and Purposes in the Morning will set a Guard upon the Soul, and fortify it under all the Temptations of the Day.

But such Self-inspection, however, should not fail to make Part of our *Evening* Devotions. When we should review and examine the several Actions of the Day, the various Tempers and Dispositions we have been in, and the Occasions that excited them. It is an Advice worthy of a Christian, tho' it first dropped from a Heathen Pen; that before we betake ourselves to Rest, we review and examine all the Passages of the Day, that we may have the Comfort of what we have done aright, and may redress what we find to have been amiss; and make the Shipwrecks of one Day be as Marks to direct our Course on another. A Practice that hath been recommended by many of the Heathen Moralists of the greatest Name, as *Plutarch*, *Epictetus*, *Marcus Antoninus*; and particularly *Pythagoras*, in the Verses that go under his Name, and are called his *Golden Verses*. Wherein he advises his Scholars every Night to recollect the Passages of the Day, and ask themselves these Questions; " Wherein have I " transgressed this Day? What have I done? " What Duty have I omitted, &c." (*h*)? *Seneca* recom-

---

(*b*) Μηδ' ὕπνον μαλακοῖσιν ἐπ' ὄμμασι προσδέξασθαι,
Πρὶν τῶν ἡμερινῶν ἔργων τρὶς ἕκαστον ἐπελθεῖν·
Πᾶ παρέβην; τί δ' ἔρεξα; τί μοι δέον οὐκ ἐτελέσθη;
Ἀρξάμενος

recommends the same Practice. 'Sextius (saith he) did this; at the Close of the Day, before he betook himself to rest, he addressed his Soul in the following Manner.' "What Evil of thine hast thou cured this Day? What Vice withstood? In what Respect art thou better?" 'Passion will cease, or become more cool, when it knows every Day it is to be thus called to Account. What can be more advantageous than this constant Custom of Searching through the Day? ——— And the same Course (saith *Seneca*) I take myself; and every Day-sit in Judgment on myself; and at Even, when all is hush and still, I make a Scrutiny into the Day; look over my Words and Actions, and hide nothing from myself; conceal none of my Mistakes through Fear; for why should I? When I have it in my Power to say thus; " This once I forgive thee; but

Αρξαμενος δ' απο πρωτε, επεξιθι ναι μεταπιςα,
Δειλα μην εκπρηξας, επιπλησσο· χρηςα δε τερπε.
Ταυτα πονει, ταυτ' εκμελετα· τετων χρη εραν σε·
Ταυτα σε της θειης αρετης εις ιχνια θησει·

Vid. Pythag. Aur. Carm. apud Poet. Minor. pag. 420.

Let not your Eyes the Sweets of Slumber taste,
Till you have thrice severe Reflections past
On th' Actions of the Day from first to last.
Wherein have I transgress'd? What done have I?
What Actions unperform'd have I pass by?
And if your Actions ill, on Search you find;
Let Grief; if good, let Joy possess your Mind.
This do, this think, to this your Heart incline,
This Way will lead you to the Life divine.

"see thou do so no more. —— In such a Dispute I was too keen; do not for the future contend with ignorant Men; they will not be convinced, because they are unwilling to shew their Ignorance.————Such a one I reproved with too much Freedom; whereby I have not reformed, but exasperated him; remember hereafter to be more mild in your Censures; and consider not only whether what you say be true, but whether the Person you say it to can bear to hear the Truth (*i*)."———Thus far that excellent Moralist.

Let us take a few other Specimens of a more pious and Christian Turn, from a judicious and devout Writer. (*k*).

' This Morning, when I arose, instead of ap-
' plying myself to GOD in Prayer, (which I
' generally find it best to do, immediately after
' a few serious Reflections) I gave way to idle
' Musing, to the great Disorder of my Heart
' and Frame. How often have I suffered for
' Want of more Watchfulness on this Occa-
' sion? When shall I be wise?——— I have this
' Day shamefully trifled, almost through the
' Whole of it: Was in my Bed when I should
' have been upon my Knees; prayed but cooly
' in the Morning; was strangely off my Guard
' in the Business and Conversation I was con-
' cerned with in the Day, particularly at ———;
  ' I in-

(*i*) *Vid. Seneca de Irâ, lib.* 3. *cap.* 36.
(*k*) M. *Bennet. See his Christ. Orator. pag.* 584.

'I indulged to very foolish, sinful, vile Thoughts, &c. I fell in with a Strain of Conversation too common amongst all Sorts, viz. *speaking Evil of others; taking up a Reproach against my Neighbour.* I have often resolved against this Sin, and yet run into it again. How treacherous this wicked Heart of mine! I have lost several Hours this Day in mere Sauntering and Idleness.——— This Day I had an Instance of mine own Infirmity, that I was a little surprised at, and I am sure I ought to be humbled for. The Behaviour of———, from whom I can expect nothing but Humour, Indiscretion, and Folly, strangely ruffled me; and that after I have had Warning over and over again. What a poor, impotent, contemptible Creature am I!——— This Day I have been kept in a great Measure from my too frequent Failings.——— I had this Day very comfortable Assistances from GOD, upon an Occasion not a little trying———what shall I render?'———

(3.) See that the Mind be in the most *composed* and *disengaged* Frame it can, when you enter upon this Business of *Self-Judgment*. Chuse a Time when it is most free from Passion, and most at Leisure from the Cares and Affairs of Life. A Judge is not like to bring a Cause to a good Issue, that is either intoxicated with Liquor on the Bench, or has his Mind distracted with other Cares, when he should be intent on the Trial. Remember you sit in Judgment upon yourself, and have nothing to do at present but to sift the Evidence which Conscience may bring in either for or against you, in order to pronounce a just Sentence; which is of much greater Concernment to you at present than any Thing else can be: And therefore it should be transacted with the utmost Care, Composure, and Attention.

(4.) Beware of *Partiality*, and the Influence of Self-love in this weighty Business; which if you do not guard against, it will soon lead you into Self-delusion; the Consequences of which may be fatal to you. Labour to see yourself as you are; and view Things in a just Light, and not in that in which you would have them appear. Remember that the Mind is always apt to believe those Things which it would have to be true, and backward to credit what it wishes to be false; and this is an Influence you will certainly lie under in this Affair of Self-judgment.

You need not be much afraid of being too severe upon yourself. Your great Danger will generally be of passing a too favourable Judgment. A Judge ought not indeed to be a Party concerned; and should have no Interest in the Person he sits in Judgment upon. But this cannot be the Case here; as you yourself are both Judge and Criminal. Which shews the Danger of pronouncing a too favourable Sentence. But remember, 'your Business is only with the

Evidence

Evidence and the Rule of Judgment; and that, however you come off now, there will be a Re-hearing in another Court, where Judgment will be according to Truth.

'However, look not unequally either at the
'Good or Evil that is in you; but view them as
'they are. If you obferve only the Good that is
'in you, and overlook the Bad, or fearch only af-
'ter your Faults, and overlook your Graces, nei-
'ther of thefe will bring you to a true Acquaint-
'ance with yourfelf (*l*).'

And to induce you to this Impartiality, remember that this Bufinefs (tho' it may be hid from the World) is not done in Secret; GOD fees how you manage it, before whofe Tribunal you muft expect a righteous Judgment. "We fhould or-
"der our Thoughts fo (faith *Seneca*) as if we had
"a Window in our Breafts, thro' which any one
"might fee what paffes there. And indeed there is
"one that does; for what does it fignify that our
"Thoughts are hid from Men? From GOD no-
"thing is hid (*m*)."

(5.) Beware of *falfe Rules* of Judgment. This is a fure and common Way to Self-deception, *e. g.* Some judge of themfelves by *what they have been*. But it does not follow, if Men are not fo bad as they have been, that therefore

---

(*l*) Baxter's *Director*, *pag.* 876.

(*m*) Sic cogitandum tanquam aliquis in pectus intimum infpicere poffit; et poteft. Quid enim prodeft ad homine aliquid effe fecretum? Nihil Deo claufum eft. *Sen. Epift.* 84.

fore they are as good as they should be. It is wrong to make our past Conduct implicitly the Measure of our present; or the present the Rule of our future; when our past, present, and future Conduct must be all brought to another Rule. And they who thus *measure themselves by themselves, and compare themselves with themselves, are not wise* *.—Again, others are apt to judge of themselves by the *Opinions of Men*; which is the most uncertain Rule that can be; for in that very Opinion of theirs you may be deceived. How do you know they have really formed so good an Idea of you as they profess? But if they have; may not others have formed as bad? And why should not the Judgment of *these* be your Rule, as well as the Opinion of those? Appeal to Self-flattery for an Answer.————However, neither one nor the other of them perhaps appear even to know *themselves*; and how should they know you? How is it possible they should have Opportunities of knowing you better than you know yourself? A Man can never gain a right Knowledge of himself from the Opinion of others which is so various, and generally so ill-founded. For Men commonly judge by outward Appearances, or inward Prejudice, and therefore for the most Part think and speak of us very much at Random. ————Again, others are for judging of themselves by the *Conduct of their Superiors*, who have Opportunities and Advantages of knowing, acting,

and

---

* 2 *Cor.* x. 12.

and being better; 'and yet without Vanity be it
'spoken (say they) *we are not behind-hand with
'them.'* But what then? Neither they nor you
perhaps are what the Obligations of your Character indispensibly require you to be, and
what you must be e'er you can be happy.
But consider how easily this Argument may be
retorted. You are better than some, you say,
who have greater Opportunities and Advantages
of being good than you have; and therefore
your State is safe. But you yourself have greater
Opportunities and Advantages of being good than
some others have, who are nevertheless better
than you; and therefore, by the same Rule, your
State cannot be safe.——Again, others judge of
themselves by the *common Maxims* of the vulgar
World concerning Honour and Honesty, Virtue
and Interest; which Maxims, tho' generally very
corrupt and very contrary to those of Reason, Conscience, and Scripture, Men will follow as a Rule,
for the Sake of the Latitude it allows them: And
fondly think, that if they stand right in the Opinion of the lowest Kind of Men, they have no
Reason to be severe upon themselves. Others,
whose Sentiments are more delicate and refined,
they imagine, may be mistaken, or may overstrain
the Matter. In which Persuasion they are confirmed, by observing how seldom the Consciences
of the Generality of Men smite them for those
Things which these nice Judges condemn as hei-
nous Crimes. I need not say how false and pernicious

nicious a Rule this is.———Again, others may judge of themselves and their State by *sudden Impressions* they have had, or strong Impulses upon their Spirits, which they attribute to the Finger of GOD; and by which they have been so exceedingly affected as to make no Doubt but that it was the Instant of their Conversion. But whether it was or no, can never be known but by the Conduct of their After-lives.——In like manner, others judge of their good State by their good *Frames*; tho' very rare, it may be, and very transient; soon passing off *like a Morning Cloud, or as the early Dew.* ' But we should not judge of ' ourselves, by that which is unusual or extraor- ' dinary with us; but by the ordinary Tenor and ' Drift of our Lives. A bad Man may seem good ' in some good Mood; and a good Man may ' seem bad in some extraordinary Falls; to judge ' of a bad Man by his best Hours, and a good ' Man by his worst, is the Way to be deceived ' in them both (*n*).' And the same Way may you be deceived in yourself.—— *Pharaoh, Ahab, Herod* and *Felix*, had all of them their Softenings, their transitory Fits of Goodness; but yet they remain upon Record under the blackest Characters.

These then are all wrong Rules of Judgment; and to trust to them, or to try ourselves by them, leads to fatal Self-deception. Again,

(6.) In

---

* Baxter's *Direct.* pag. 876.

(6.) In the Business of Self-examination you must not only take Care you do not judge by wrong Rules, but that you do not judge *wrong by right Rules.* You must endeavour then to be well acquainted with them. The Office of a Judge is not only to collect the Evidence and the Circumstances of Facts, but to be well skilled in the Laws by which those Facts are to be examined.

Now the only right Rules by which we are to examine, in order to know ourselves, are *Reason* and *Scripture.* Some are for setting aside these Rules, as too severe for them; too stiff to bend to their Perverseness; too streight to measure their crooked Ways! are against Reason, when Reason is against them; decrying it as *carnal Reason:* And against Scripture, when Scripture is against them, despising it as a *dead Letter.* And thus, rather than be convinced they are wrong, they reject the only Means that can set them right.

And as some are for setting aside these Rules, so others are for setting them one against the other. Reason against Scripture, and Scripture against Reason. When they are both given us by the GOD of our Natures, not only as perfectly consistent, but as proper to explain and illustrate each other, and prevent our mistaking either; and to be, when taken together, (as they always should) the most complete and

only

only Rule by which to judge both of ourselves, and every Thing belonging to our Salvation, as reasonable and fallen Creatures.

(1.) Then one Part of that Rule which GOD hath given us to judge of ourselves by, is *right Reason*. By which I do not mean the Reasoning of any particular Man, which may be very different from the Reasoning of another particular Man; and both, it may be, very different from *right Reason*; because both may be influenced not so much by the Reason and Nature of Things, as by partial Prepossessions and the Power of Passions. But by *right Reason* I mean those common Principles, which are readily allowed by all who are capable of understanding them, and not notoriously perverted by the Force of Prejudice; and which are confirmed by the common Consent of all the sober and thinking Part of Mankind; and may be easily learned by the Light of Nature. Therefore if any Doctrine or Practice, tho' supposed to be founded in, or countenanced by Revelation, be nevertheless apparently repugnant to these Dictates of right Reason, or evidently contradict our natural Notions of the Divine Attributes, or weaken our Obligations to universal Virtue, *that* we may be sure is no Part of Revelation; because then one Part of our Rule would clash with and be opposite to the other. And thus *Reason* was designed to be our Guard against a wild and extravagant Construction of *Scripture*.

(2.) The

(2.) The other Part of our Rule is the *Sacred Scriptures*, which we are to use as our Guard against the licentious Excursions of *Fancy*, which is often imposing itself upon us for *right Reason*. Let any religious Scheme or Notion then appear ever so pleasing or plausible, if it be not established on the plain Principles of Scripture, it is forthwith to be discarded: and that Sense of Scripture that is violently forced to bend towards it, is very much to be suspected.

It must be very surprizing to one who reads and studies the sacred Scriptures with a free, unbiassed Mind, to see what elaborate, fine-spun flimsy Glosses Men will invent and put upon some Texts as the true and genuine Sense of them; for no other Reason, but because it is most agreeable to the Opinion of their Party, from which, as the Standard of their Orthodoxy, they durst never depart; who, if they were to write a Critique in the same Manner on any *Greek* or *Latin* Author, would make themselves extremely ridiculous in the Eyes of the learned World. But, if we would not pervert our Rule, we must learn to think as Scripture speaks, and not compel that to speak as we think.

Would we know ourselves then, we must often view ourselves in the Glass of GOD's Word. And when we have taken a full Survey of ourselves from thence, let us not soon forget *what Manner of Persons we are* [*]. If our own Image do not

please

---

[*] *Jam.* i. 23. 24.

please us, let us not quarrel with our Mirrour, but set about mending ourselves.

The Eye of the Mind indeed is not like that of the Body, which can see every Thing else but itself; for the Eye of the Mind can turn itself inward, and survey itself. However, it must be owned, it can see itself much better when its own Image is reflected upon it from this Mirrour. And it is by this only that we can come at the Bottom of our Hearts, and discover those secret Prejudices and carnal Prepossessions, which Self-love would hide from us.

This then is the first Thing we must do in order to Self-Knowledge. We must *examine*, scrutinize, and judge ourselves, *diligently*, *leisurely*, *frequently*, and *impartially*; and that not by the *false Maxims* of the World, but by the *Rules* which GOD hath given us, *Reason*, and *Scripture*; and take Care to understand those Rules, and not set them at *Variance*.

CHAP. II.

Constant Watchfulness necessary to Self-Knowledge.

II. WOULD *we know ourselves, we must be very* watchful *over our Hearts and Lives.*

(1.) We must keep a vigilant Eye upon our *Hearts, i. e.* our Tempers, Inclinations and Passions. A more necessary Piece of Advice, in order to Self-acquaintance, there cannot be, than that which *Solomon* gives us *, *Keep your Heart with all Diligence,* or as it is in the Original, *above all keeping* (*o*). *q. d.* Whatever you neglect or overlook, be sure you mind your Heart (*p*). Narrowly observe all its Inclinations and Aversions, all its Motions and Affections, together with the several Objects and Occasions which excite them. And this Precept we find in Scripture inforced with two very urgent Reasons. The first is, because *out of it are the Issues of Life, i. e.* As our Heart is, so will the Tenor of our Life and Conduct be. As is the Fountain, so are the Streams; as is the Root, so is the

---

* *Prov.* iv. 23.

(*o*) מכל־משמר

(*p*) Parallel to this Advice of the *Royal Preacher,* is that of the *Imperial Philosopher,* Ενδον βλεπε, ενδον γαρ η πηγη τε αγαθε. *Look within; for within is the Fountain of Good.* M. *Aurel.* lib. 7. § 59.

the Fruit *. And the other is, becaufe *it is deceitful above all Things* †. And therefore, without a conftant Guard upon it, we fhall infenfibly run into many hurtful Self-deceptions. To which I may add, that without this careful Keeping of the Heart, we fhall never be able to acquire any confiderable Degree of Self-acquaintance or Self-government.

(2.) To know ourfelves, we muft watch our *Life* and *Conduct* as well as our Hearts. And by this the Heart will be better known; as the Root is beft known by the Fruit. We muft attend to the Nature and Confequences of every Action we are difpofed or folicited to, before we comply; and confider how it will appear in a future Review. We are apt enough to obferve and watch the Conduct of others: A wife Man will be as critical and as fevere upon his own. For indeed we have a great deal more to do with our own Conduct than that of other Men; as we are to anfwer for our own, but not for theirs. By obferving the Conduct of other Men we know *them*, by carefully obferving our own, we muft know *ourfelves*.

* *Mat.* vii. 18.   † *Jer.* xvii. 9.

CHAP. III.

We should have some Regard to the Opinions of others concerning us, particularly of our Enemies.

III. WOULD we know ourselves, we should not altogether neglect the Opinion which others may entertain concerning us.

Not that we need be very solicitous about the Censure or Applause of the World; which are generally very rash and wrong, and proceed from the particular Humours and Prepossessions of Men: and he that knows himself, will soon know how to despise them both. 'The Judgment which 'the World makes of us, is generally of no Man-'ner of Use to us; it adds nothing to our Souls 'or Bodies, nor lessens any of our Miseries. Let 'us constantly follow Reason, (says *Montaigne*) 'and let the publick Approbation follow us the 'same Way, if it pleases.'

But still, I say, a total Indifference in this Matter is unwise (*q*). We ought not to be intirely

---

(*q*) Τὰς δὲ ἀκαίρους κατηγορίας—ἐδὲ ἀμέτρως δεδοικέναι καὶ τρέμειν ἐδὲ ἁπλῶς παρορᾶν καλόν· ἀλλὰ, χρὴ κἂν ψυδεῖς, τυγχάνωσιν, ἔσαι, κἂν παρὰ τῶν τυχόντων ἡμῖν, ἐπάγωνται, πειρᾶσθαι σβιννύναι ταχέως αὐτάς. *Chrysost. de Sacred. l.* 5. *c.* 4. As to the groundless Reports that may be raised to our Disadvantage, it is not good either too much to fear them, or entirely to despise them. We should endeavour to stifle them, be they ever so false, or the Authors of them ever so contemptible.

tirely infensible of the Reports of others; no, not to the Railings of an Enemy; for an Enemy may say something out of Ill-will to us, which it may concern us to think of coolly when we are by ourselves; to examine whether the Accusation be just; and what there is in our Conduct and Temper which may make it appear so. And by this Means our Enemy may do us more Good than he intended; and discover to us something in our Hearts which we did not before advert to. A Man that hath no Enemies ought to have very faithful Friends; and one who hath no such Friends, ought to think it no Calamity that he hath Enemies to be his effectual Monitors. ———— 'Our Friends (says
' Mr. *Addison*) very often flatter us as much as
' our own Hearts. They either do not see our
' Faults, or conceal them from us; or soften them
' by their Representations, after such a Manner
' that we think them too trivial to be taken Notice
' of. An Adversary, on the contrary, makes a
' stricter Search into us, discovers every Flaw and
' Imperfection in our Tempers, and though his
' Malice may set them in too strong a Light, it
' has generally some Ground for what it advances.
' A Friend exaggerates a Man's Virtues, an Enemy
' my inflames his Crimes. A wise Man should
' give a just Attention to both of them, so far
' as it may tend to the Improvement of the one,
' and the Diminution of the other. *Plutarch*
' has written an Essay on the Benefits which a

' Man may receive from his Enemies; and a-
' mong the good Fruits of Enmity mentions this
' in particular, that by the Reproaches it casts up-
' on us we see the worst Side of ourselves, and o-
' pen our Eyes to several Blemishes and Defects in
' our Lives and Conversations, which we should
' not have observed, without the Help of such ill-
' natured Monitors.

' In order likewise to come at a true Knowledge
' of ourselves, we should consider, on the other
' hand, how far we may deserve the Praises and
' Approbations which the World bestow upon us;
' whether the Actions they celebrate proceed from
' laudable and worthy Motives, and how far we
' are really possessed of the Virtues which gain us
' Applause amongst those with whom we converse.
' Such a Reflection is absolutely necessary, if we
' consider how apt we are either to value or con-
' demn ourselves by the Opinions of others, and
' to sacrifice the Report of our own Hearts to the
' Judgment of the World (r).'

In that Treatise of *Plutarch* here referred to, there are a great many excellent Things pertinent to this Subject; and therefore I thought it not improper to throw a few Extracts out of it into the Margin (s).

It

(r) *Spectat. Vol.* vi. *No.* 399.

(s) The Foolish and Inconsiderate spoil the very Friendships they are ingaged in; but the Wise and Prudent make good Use of the Hatred and Enmity of Men against them.

Why should we not take an Enemy for our Tutor, who will instruct us *gratis* in those Things we knew not before? For an Enemy,

It is the Character of a dissolute Mind, to be entirely insensible to all that the World says of us; and shews such a Confidence of Self-Knowledge, as is usually a sure Sign of *Self-ignorance*. The most knowing Minds are ever least presumptuous. And true Self-Knowledge is a Science of so much Depth and Difficulty, that a wise Man would

Enemy sees and understands more in Matters relating to us than our Friends do. Because Love is blind, but Spite, Malice, Ill-will, Wrath, and Contempt, talk much, are very inquisitive and quick-sighted.

Our Enemy, to gratify his Ill-will towards us, acquaints himself with the Infirmities both of our Bodies and Minds; sticks to our Faults, and makes his invidious Remarks upon them, and spreads them abroad by his uncharitable and ill-natured Reports. Hence we are taught this useful Lesson for the Direction and Management of our Conversation in the World, *viz.* that we be circumspect and wary in every Thing we speak or do, as if our Enemy always stood at our Elbow, and overlook'd our Actions.

Those Persons whom *that* Wisdom hath brought to live soberly, which the Fear and Awe of Enemies hath infused, are by Degrees drawn into a Habit of living so, and are composed and fixed in their Obedience to Virtue by Custom and Use.

When one asked *Diogenes* how he might be avenged of his Enemies, he replied, *To be yourself a good and honest Man.*

Antisthenes spake incomparably well; "that, if a Man would
" live a safe and unblameable Life, it was necessary that he should
" have very ingenuous and faithful Friends, or every bad Enemies;
" because the first, by their kind Admonitions, would keep him
" from sinning, the latter by their Invectives."

He that hath no Friend to give him Advice, or reprove him when he does amiss, must bear patiently the Rebukes of his Enemies, and thereby learn to mend the Errors of his Ways; considering seriously the Object which these severe Censures aim at, and not what he his who makes them. For he who designed the Death of *Prometheus* the *Thessalian*, instead of giving him a fatal Blow, only opened

would not chuse to be over-confident that all his Notions of himself are right, in Opposition to the Judgment of all Mankind; some of whom perhaps have better Opportunities and Advantages of knowing him (at some Seasons especially) than he has of knowing himself. Because herein they never look through the same false Medium of *Self-flattery*.

CHAP.

opened a Swelling which he had, which did really save his Life. Just so may the harsh Reprehensions of Enemies cure some Distempers of the Mind, which were before either not known or neglected; though their angry Speeches do originally proceed from Malice or Ill-will.

If any Man with opprobrious Language objects to you Crimes you know nothing of, you ought to enquire into the Causes or Reasons of such false Accusations; whereby you may learn to take Heed for the future, lest you should unwarily commit those Offences which are unjustly imputed to you.

Whenever any Thing is spoken against you that is not true, do not pass by, or despise it because it is false; but forthwith examine yourself, and consider what you have said or done that may administer a just Occasion of Reproof.

Nothing can be a greater Instance of Wisdom and Humanity, than for a Man to bear silently and quietly the Follies and Revilings of an Enemy; taking as much Care not to provoke him, as he would to sail safely by a dangerous Rock.

It is an eminent Piece of Humanity, and a manifest Token of a Nature truly generous, to put up the Affronts of an Enemy, at a Time when you have a fair Opportunity to revenge them.

Let us carefully observe those good Qualities wherein our Enemies excel us. And endeavour to excel them, by avoiding what is faulty, and imitating what is excellent in them. *Plut. Mor. Vol.* i. *pag.* 265. *et seq.*

CHAP. IV.

Frequent Converse with Superiors a Help to Self-Knowledge.

IV. ANOTHER *proper Means of Self-Knowledge, is to converse as much as you can with those who are your Superiors in real Excellence.*

He that walketh with wise Men shall be wise *. Their Example will not only be your Motive to laudable Pursuits, but a Mirrour to your Mind; by which you may possibly discern some Failings or Deficiencies or Neglects in yourself, which before escaped you. You will see the Unreasonableness of your Vanity and Self-sufficiency, when you observe how much you are surpassed by others in Knowledge and Goodness. Their Proficiency will make your Defects the more obvious to yourself. And by the Lustre of their Virtues you will better see the Deformity of your Vices; your Negligence by their Diligence; your Pride by their Humility; your Passion by their Meekness, and your Folly by their Wisdom.

Examples not only move, but teach and direct, much more effectually than Precepts; and shew us not only that such Virtues may be practised, but *how*; and how lovely they appear when

* *Prov.* xiii. 20.

when they are. And therefore, if we cannot have them always before our Eyes, we should endeavour to have them always in our Mind; and especially that of our great Head and Pattern, who hath set us a perfect Example of the most innocent Conduct under the worst and most disadvantageous Circumstances of Human Life (*t*).

CHAP. V.

Of cultivating such a Temper as will be the best Disposition to Self-Knowledge.

V. IF a Man would know himself, he must with great Care cultivate that Temper which will best dispose him to receive this Knowledge.

Now as there are no greater Hindrances to Self-Knowledge than Pride and Obstinacy; so there is nothing more helpful to it than *Humility* and an *Openness to Conviction.*

1. One who is in Quest of *Self-Knowledge*, must above all Things seek *Humility*. And how near an Affinity there is between these two appears from hence, that they are both acquired the same Way. The very Means of attaining Humility are the properest Means for attaining Self-

---

(*t*) Qui plenissimè intelligere appetit qualis sit, tales debet aspicere qualis non est; ut in bonorum formâ, metiatur quantum deformis est. *Greg.*

Self-acquaintance. By keeping an Eye every Day upon our Faults and Wants we become more humble; and by the same Means we become more Self-intelligent. By considering how far we fall short of our Rule and our Duty, and how vastly others exceed us, and especially by a daily and diligent Study of the Word of GOD, we come to have meaner Thoughts of ourselves; and by the very same Means we come to have a better Acquaintance with ourselves.

A proud Man cannot know himself. Pride is that *Beam* in the Eye of his Mind, which renders him quite blind to any Blemishes there. Hence nothing is a surer Sign of Self-ignorance than Vanity and Ostentation.

Indeed true Self-Knowledge and Humility are so necessarily connected, that they depend upon, and mutually beget each other. A Man that knows himself knows the Worst of himself, and therefore cannot but be humble; and a humble Mind is frequently contemplating its own Faults and Weaknesses, which greatly improves it in Self-Knowledge. So that Self-acquaintance makes a Man humble; and Humility gives him still a better Acquaintance with himself.

(2.) An *Openness* to *Conviction* is no less necessary to Self knowledge than Humility.

As nothing is a greater Bar to true Knowledge than an obstinate Stiffness in Opinion, and a Fear to depart from old Notions, which (before we were capable of judging perhaps) we had long taken

taken up for the Truth; so nothing is a greater Bar to Self-Knowledge, than a strong Aversion to part with those Sentiments of *ourselves* which we have been blindly accustomed to, and to think worse of ourselves than we are wont to do.

And such an Unwillingness to retract our Sentiments in both Cases proceeds from the same Cause, *viz.* a Reluctance to Self-condemnation. For he that takes up a new Way of thinking, contrary to that which he hath long received, therein condemns himself of having lived in an Error; and he that begins to see Faults in himself he never saw before, condemns himself of having lived in Ignorance and Sin. Now this is a most ungrateful Business, and what Self-flattery can by no Means endure.

But such an Inflexibility of Judgment, and Hatred of Conviction, is a very unhappy and hurtful Turn of Mind. And a Man that is resolved never to be in the Wrong, is in a fair Way never to be in the Right.

As Infallibility is no Privilege of the Human Nature, it is no Diminution to a Man's good Sense or Judgment to be found in an Error, provided he is willing to retract it. He acts with the same Freedom and Liberty as before, whoever be his Monitor; and it is his own good Sense and Judgment that still guides him; which shines to great Advantage in thus directing him against the Bias of Vanity and Self-opinion. And in thus changing his Sentiments, he only acknowledges that

that he is not (what no Man ever was) incapable of being mistaken. In short it is more Merit, and an Argument of a more excellent Mind, for a Man freely to retract when he is in the Wrong, than to be overbearing and positive when he is in the Right (*u*).

A Man then must be *willing* to know himself, before he *can* know himself. He must open his Eyes, if he desires to see; yield to Evidence and Conviction, though it be at the Expence of his Judgment, and to the Mortification of his Vanity.

CHAP. VI.

To be sensible of our false Knowledge, a good Step to Self-Knowledge.

VI. WOULD *you know yourself, take Heed and guard against false Knowledge.*
See that the *Light that is within you be not Darkness*; that your favourite and leading Principles be right. Search your Furniture, and consider what you have to unlearn. For oftentimes there

---

(*u*) Ει τις με ελεγξαι, και παραστησαι μοι, ότι ουκ ορθως υπολαμβανω η πραττω, δυναται, χαιρων μεταθησομαι· ζητω γαρ την αληθειαν υφ ης ουδεις πωποτε εβλαβη· βλαπτεται δε ο επιμενων επι της εαυτου απατης και αγνοιας. *M. Aur. lib.* 6. § 21.
If any one can convince me that I am wrong in any Point of Sentiment or Practice, I will alter it with all my Heart. For it is Truth I seek; and that can hurt no Body. It is only persisting in Error or Ignorance that can hurt us.

there is as much Wisdom in casting off some Knowledge which we have, as in acquiring that which we have not. Which perhaps was what made *Themistocles* reply, when one offered to teach him the Art of Memory, that *he had much rather he would teach him the Art of Forgetfulness.*

A Scholar that hath been all his Life collecting Books, will find in his Library at last a great deal of Rubbish. And as his Taste alters, and his Judgment improves, he will throw out a great many as Trash and Lumber, which, it may be, he once valued and paid dear for; and replace them with such as are more solid and useful. Just so should we deal with our Understandings; look over the Furniture of the Mind; separate the Chaff from the Wheat, which are generally received into it together; and take as much Pains to forget what we ought not to have learned, as to retain what we ought not to forget. To read Froth and Trifles all our Life, is the Way always to retain a flashy and juvenile Turn; and only to contemplate our first (which is generally our worst) Knowledge, cramps the Progress of the Understanding, and makes our Self-survey extremely deficient. In short, would we improve the Understanding to the valuable Purposes of Self-Knowledge, we must take as much Care what Books we read, as what Company we keep.

" The Pains we take in Books or Arts, which
" treat of Things remote from the Use of Life,
" is a busy Idleness. If I study (says *Montaigne*)

"it is for no other Science than what treats of the Knowledge of myself, and instructs me how to live and die well (x)."

It is a comfortless Speculation, and a plain Proof of the Imperfection of the Human Understanding, that upon a narrow Scrutiny into our Furniture, we observe a great many Things which we think we know, but do not; and many which we do know, but ought not; that a good deal of the Knowledge we have been all our Lives collecting, is no better than mere Ignorance, and some of it worse; to be sensible of which is a very necessary Step to Self-acquaintance (y).

CHAP. VII.

Self-inspection peculiarly necessary upon some particular Occasions.

VII. WOULD *you know yourself, you must very carefully attend to the Frame and Emotions of your Mind under some extraordinary Incidents.*

Some sudden Accidents which befal you when the Mind is most off its Guard, will better discover its secret Turn and prevailing Disposition than much greater Events you are prepared to meet, *e. g.*

(1.) Consider how you behave under any sudden *Affronts* or *Provocations* from Men. A Fool's *Wrath*

(x) *Rule of Life*, pag. 82, 90.
(y) See Part i. Chap. xiii. *fin.*

Wrath is presently known *, *i. e.* a Fool is presently known by his Wrath.

If your Anger be soon kindled, it is a Sign that secret Pride lies lurking in the Heart; which, like Gun-powder, takes Fire at every Spark of Provocation that lights upon it. For whatever may be owing to a natural Temper, it is certain that Pride is the chief Cause of frequent and wrathful Resentments. For Pride and Anger are as nearly allied as Humility and Meekness. *Only by Pride cometh Contention* †. And a Man would not know what Mud lay at the Bottom of his Heart, if Provocation did not stir it up.

Athenodorus the Philosopher, by Reason of his old Age, begged leave to retire from the Court of *Augustus*, which the Emperor granted him; and in his Compliments of Leave, 'Remember *(said he)* ' *Cæsar*, whenever you are angry, you say or do ' nothing, before you have distinctly repeated to ' yourself the four and twenty Letters of the Al-' phabet.' Whereupon *Cæsar* catching him by the Hand, *I have need* (says he) *of your Presence still*; and kept him a Year longer *(z)*. This is celebrated by the Antients as a Rule of excellent Wisdom. But a Christian may prescribe to himself a much wiser, *viz.* ' When you are angry, ' answer not till you have repeated the fifth Petiti-' on of the Lord's Prayer, *forgive us our Tres-' passes, as we forgive them that trespass against us.*
' And

---

* *Prov.* xii. 16. † *Prov.* xiii. 10.
(*z*) See *Plut. Mor. Vol.* i. *pag.* 238.

'And our Saviour's Comment upon it.' *For if ye forgive Men their Trespasses, your heavenly Father will also forgive you: But if ye forgive not Men their Trespasses, neither will your Father forgive your Trespasses* *.

It is a just and seasonable Thought, that of *Marcus Antoninus* upon such Occasions; "A Man misbehaves himself towards me,——what is that to me? The Action is his; and the Will that sets him upon it is his; and therefore let him look to it. The Fault and Injury belong to him, not to me. As for me, I am in the Condition Providence would have me, and am doing what becomes me (*a*)."

But after all, this amounts only to a philosophical Contempt of Injuries; and falls much beneath the Dignity of a Christian Forgiveness, to which Self Knowledge will happily dispose us. And therefore, in order to judge of our Improvements therein, we must always take Care to examine and observe, in what Manner we are affected in such Circumstances.

(2.) How do you behave under a severe and unexpected *Affliction* from the *Hand of Providence?* Which is another Circumstance, wherein we have a fair Opportunity of coming to a right Knowledge of ourselves.

If there be an habitual Discontent or Impatience lurking within us, this will draw it forth. Especially if the Affliction be attended with any
of

---

* *Mat.* vi. 14, 15.
(*a*) *Meditat. Book* 5. § 25.

of those aggravating Circumstances which accumulated that of *Job*.

Afflictions are often sent with this Intent, to teach us to know ourselves; and therefore ought to be carefully improved to this Purpose.

And much of the Wisdom and Goodness of our heavenly Father is seen by a serious and attentive Mind, nor only in proportioning the Degrees of his Corrections to his Children's Strength, but in adapting the Kinds of them to their Tempers; afflicting one in one Way, another in another, according as he knows they are most easily wrought upon, and as will be most for their Advantage. By which Means a small Affliction of one Kind may as deeply affect us, and be of more Advantage to us, than a much greater of another.

It is a trite but true Observation, that a wise Man receives more Benefit from his Enemies, than from his Friends; from his Afflictions than from his Mercies; by which Means his Enemies become in Effect his best Friends, and his Afflictions his greatest Mercies. Certain it is, that a Man never has an Opportunity of taking a more fair and undisguised View of himself, than in these Circumstances. And therefore by diligently observing in what Manner he is affected at such Times, he may make an Improvement in the true Knowledge of himself, very much to his future Advantage, tho' perhaps not a little to his present Mortification. For a sudden Provocation from Man,

Man, or a severe Affliction from GOD, may detect something which lay latent and undiscovered so long at the Bottom of his Heart, that he never once suspected it to have had any Place there. Thus the one excited Wrath in the *meekest* Man *, and the other Passion in the most *patient* †.

By considering then in what Manner we bear the particular Afflictions GOD is pleased to allot us, and what Benefit we receive from them, we may come to a very considerable Acquaintance with ourselves.

(3.) What is our usual Temper and Disposition in a Time of *Peace, Prosperity* and *Pleasure,* when the Soul is generally most unguarded?

This is the warm Season that nourishes and impregnates the Seeds of Vanity, Self-confidence, and a supercilious Contempt of others. If there be such a *Root of Bitterness* in the Heart, it will be very apt to shoot forth in the Sunshine of uninterrupted Prosperity; even after the Frost of Adversity had nipped it, and, as we thought, killed it.

Prosperity is a Trial as well as Adversity; and is commonly attended with more dangerous Temptations. And were the Mind but as seriously disposed to Self-reflection, it would have a greater Advantage of attaining a true Knowledge of itself under the former than under the latter. But the Unhappiness of it is, the Mind is seldom rightly turned for such an Employment under those Circumstances.

* *Psal.* cvi. 33.   † *Job* iii. 3.

cumstances. It has something else to do; has the Concerns of the World to mind; and is too much engaged by the Things without it, to advert to those within; and is more disposed to enjoy than *examine* itself. However, it is a very necessary Season for Self-examination, and a very proper Time to acquire a good Degree of Self-acquaintance, if rightly improved.

(*Lastly,*) How do we behave in *bad Company* ?

And that is to be reckoned bad Company in which there is no Probability of our doing or getting any Good, but apparent Danger of our doing or getting much Harm ; I mean, our giving Offence to others, by an indiscreet Zeal, or incurring Guilt to ourselves by a criminal Compliance.

Are we carried down by the Torrent of Vanity and Vice? Will a Flash of Wit or a brilliant Fancy make us excuse a profane Expression? If so, we shall soon come to relish it, when thus seasoned, and use it ourselves.

This is a Time when our Zeal and Wisdom, our Fortitude and Firmness are generally put to the most delicate Proof; and when we may too often take Notice of the unsuspected Escapes of Folly, Fickleness and Indiscretion.

At such Seasons as these then, we may often discern what lies at the Bottom of our Hearts, better than we can in the more even and customary Scenes of Life, when the Passions are all calm and still. And therefore would we know ourselves, we should be very attentive to our Frame, Temper,

per, Disposition, and Conduct upon such Occasions.

CHAP. VIII.

To know ourselves, we must wholly abstract from external Appearances.

VIII. WOULD *you know yourself, you must, as far as possible, get above the Influence of* Exteriors, *or a mere outward Show.*

A Man is, what his Heart is. The Knowledge of himself is the Knowledge of his Heart, which is intirely an inward Thing; to the Knowledge of which then, outward Things (such as a Man's Condition and State in the World) can contribute nothing: But on the other hand, is too often a great Bar and Hindrance to him in his Pursuit of Self-Knowledge.

(1.) Are your Circumstances in the World *easy* and *prosperous,* take care you do not judge of yourself too favourably on that Account.

These Things are without you, and therefore can never be the Measure of what is within; and however the World may respect you for them, they do not in the least make you either a wiser or more valuable Man.

In forming a true Judgment of yourself then, you must intirely set aside the Consideration of your Estate, and Family; your Wit, Beauty, Genius,

Genius, Health, &c. which are all but the Appendages or Trappings of a Man; a smooth and shining Varnish, which may lacker over the basest Metal (*b*).

A Man may be a good and happy Man without these Things, and a bad and wretched one with them. Nay he may have all these, and be the worse for them. They are so far from being good and excellent in themselves, that we often see Providence bestows them upon the vilest of Men, and in Kindness denies them to some of the best. They are oftentimes the greatest Temptations, and put a Man's Faith and Wisdom to the most dangerous Trial.

(2.) Is your Condition in Life *mean* and *afflicted?* Do not judge the worse of yourself for not having those external Advantages which others have.

None will think the worse of you for the Want of them, but those who think the better of themselves for having them: In both which they shew a very depraved and perverted Judgment. These are (τὰ ἐκ ἐφ' ἡμῖν) Things intirely *without us* and out of our Power; for which a Man is neither the better nor the worse, but according as he uses them: And therefore you ought to be as indifferent to them as they are to you. A good Man shines amiably through all the Obscurity of his

---

(*b*) Si perpendere te voles, sepone pecuniam, domum, Dignitatem, intus te ipse consule. *Sen.*

Nam genus, et proavos, et que non fecimus ipsi,
Vix ea nostra voco.   *Ovid. Met. lib.* xiii. 140.

his low Fortune; and a wicked Man is a poor little Wretch in the Midſt of all his Grandeur (*c*).

Were we to follow the Judgment of the World, we ſhould indeed think otherwiſe of theſe Things; and by that Miſtake be led into a wrong Notion of ourſelves. But we have a better Rule to follow, to which if we adhere, the Conſideration of our external Condition in Life, whatever it be, will have no undue Influence on the Mind in its Search after Self-Knowledge.

CHAP. IX.

The Practice of Self-Knowledge, a great Means to promote it.

IX. LET *all your Self-Knowledge be reduced into Practice.*

The right Improvement of that Knowledge we have, is the beſt Way to attain more.

The great End of Self-Knowledge is Self-government; without which (like all other) it is but a uſeleſs Speculation. And as all Knowledge is valuable in Proportion to its End, ſo this is the moſt excellent, only becauſe the Practice of it is of the moſt extenſive Uſe.

' Above all other Subjects (ſays an antient pious
' Writer) ſtudy thine own Self.—For no Know-
' ledge that terminates in Curioſity or Speculation
' is

(*c*) Parvos pumilio, licèt in monte conſtiterit; coloſſus magnitudinem ſuam ſervabit, etiamſi ſteterit in puteo. *Sen. Epiſt.* 77.

" Pygmies are Pygmies ſtill, tho' plac'd on Alps;
" And Pyramids are Pyramids in Vales. *Night Thoughts.*

'is comparable to that which is of use; and of all
'useful Knowledge, that is most so which consists
'in the due Care and just Notions of ourselves.
'This Study is a Debt which every one owes him-
'self. Let us not then be so lavish, so unjust as not to
'pay this Debt; by spending some Part, at least,
'if we cannot all or most of our Time and Care
'upon that which has the most indefeasible Claim
'to it. Govern your Passions; manage your Acti-
'ons with Prudence; and where false Steps have
'been made, correct them for the future. Let no-
'thing be allowed to grow headstrong and disor-
'derly; but bring all under Discipline. Set all your
'Faults before your Eyes; and pass Sentence upon
'yourself with the same Severity as you would do
'upon another, for whom no Partiality hath bi-
'assed your Judgment (*d*).'

What will our most exact and diligent Self-researches avail us, if, after all, we sink into Indolence and Sloth? Or what will it signify to be convinced that there is a great deal amiss in our Deportments and Dispositions, if we sit still contentedly under that Conviction, without taking one Step towards a Reformation? It will indeed render us but the more guilty in the Sight of GOD. And how sad a Thing will it be to have our *Self-Knowledge* hereafter rise up in Judgment against us!——

'Examination is in order to Correction and
'Amendment. We abuse it and ourselves, if
'we rest in the Duty without looking farther.
'We

(*d*) *St. Bernard's Medit. chap.* 5.

'We are to review our daily Walk, that we may
'*reform* it: and confequently a daily Review will
'point out to us the Subject and Matter of our
'future daily Care.' —— "This Day (faith the
"Chriftian upon his Review of Things at Night)
"I loft fo much Time; particularly at ———
"I took too great a Liberty; particularly in———
"I omitted fuch an Opportunity that might have
"been improved to better Purpofe. I mifmana-
"ged fuch a Duty —— I find fuch a Corruption
"often working; my old Infirmity ——— ftill
"cleaves to me: How eafily doth this Sin befet
"me!———Oh! May I be more attentive for
"the Time to come, more watchful over my
"Heart; take more Heed to my Ways! May I
"do fo the next Day!" —— 'The Know-
'ledge of a Diftemper is a good Step to a Cure;
'at leaft, it directs to proper Methods and Ap-
'plications in order to it. Self-acquaintance
'leads to Self-reformation. He that at the Clofe of
'each Day calls over what is paft, infpects himfelf,
'his Behaviour and Manners, will not fall into
'that Security, and thofe uncenfured Follies that
'are fo common and fo dangerous (*c*).'

And it may not be improper, in order to
make us fenfible of, and attentive to fome of
the more fecret Faults and Foibles of our Tempers, to pen them down at Night, according
as they appeared during the Tranfactions of

(*c*) *Bennet's Chrift. Orat. pag.* 578.

the Day. By which Means, we shall not only have a more distinct View of that Part of our Character to which we are generally most blind; but shall be able to discover some Defects and Blemishes in it, which perhaps we never apprehended before. For the Wiles and Doublings of the Heart are sometimes so hidden and intricate, that it requires the nicest Care and most steady Attention to detect and unfold them.

For Instance; 'This Day I read an Au-
'thor, whose Sentiments were very different
'from mine, and who expressed himself with
'much Warmth and Confidence. It excited
'my Spleen, I own, and I immediately passed
'a severe Censure upon him. So that had
'he been present, and talked in the same Strain,
'my ruffled Temper would have prompted me
'to use harsh and ungrateful Language, which
'might have occasioned a very unchristian
'Contention. But I now recollect, that tho'
'the Author might be mistaken in those Sen-
'timents, (as I still believe he was) yet by his
'particular Circumstances in Life, and the Me-
'thod of his Education, he hath been strongly led
'into that Way of thinking. So that his Pre-
'judice is pardonable; but my Uncharitableness
'is not; especially, considering that in many
'Respects he has the Ascendant of me.——— This
'proceeded then from *Uncharitableness*, which is
'one Fault of my Temper I have to watch a-
'gainst; and which I never was before so sensi-
'ble

'ble of, as I am now upon this Recollection.
'Learn more *Moderation*, and make more Al-
'lowances for the mistaken Opinions of others
'for the future. Be as charitable to others who
'differ from you, as you desire they should be
'to you who differ as much from them. For it
'may be you cannot be more assured of being in
'the Right than they are.

'Again; this Day I found myself strongly
'inclined to put in something by Way of A-
'batement to an excellent Character given of an
'absent Person, by one of his great Admirers.
'It is true, I had the Command of myself to
'hold my Tongue. And it is well I had; for
'the Ardour of his Zeal would not have admit-
'ted the Exception, (though I still think that in
'some Degree it was just) which might have
'raised a wrangling Debate about his Character,
'perhaps at the Expence of my own; or how-
'ever occasioned much Animosity and Conten-
'tion.————But I have since examined the
'secret Spring of that Impulse, and find it to be
'*Envy*; which I was not then sensible of; but
'my Antagonist had certainly imputed it to this.
'And had he taken the Liberty to have told me
'so, I much question whether I should have had
'the Temper of the Philosopher; who, when
'he was really injured, being asked whether he
'was angry or no, replied, *No*; *but I am consi-*
'*dering with myself whether I ought not to be so.*
'I doubt I should not have had so much Com-
'posure; but should have immediately resented

'it as a false and malicious Aspersion. But it
'was certainly Envy, and nothing else; for the
'Person who was the Object of the Encomium
'was much my Superior in many Respects. And
'the Exception that arose to my Mind was the
'only Flaw in his Character; which nothing but
'a quick-sighted Envy could descry. Take Heed
'then of that Vice for the future.

'Again; this Day I was much surprized to
'observe in myself the Symptoms of a Vice,
'which of all others, I ever thought myself
'most clear of; and have always expressed the
'greatest Detestation of in others, and that is
'*Covetousness*. For what else could it be that
'prompted me to with-hold my Charity from
'my Fellow-Creature in Distress, on Pretence
'that he was not in every Respect a proper
'Object; or to dispense it so sparingly to ano-
'ther, who I knew was so, on Pretence of
'having lately been at a considerable Expence
'upon another Occasion? This could proceed
'from nothing else but a latent Principle of
'Covetousness; which though I never before ob-
'served in myself, yet it is likely others have.
'O how inscrutable are the Depths and Deceits
'of the Human Heart!——Had my Enemy
'brought against me a Charge of Indolence,
'Self-indulgence, or Pride, and Impatience, or
'a too quick Resentment of Affronts and In-
'juries, my own Heart must have confirmed
'the Accusation, and forced me to plead guilty.
'Had he charged me with Bigotry, Self-opi-
'nion

'nion and Censoriousness, I should have thought
'it proceeded from the same Temper in *him-*
'*self*, having rarely observed any Thing like
'it in my own. But had he charged me with
'Covetousness, I should have taken it for down-
'right *Calumny*, and despised the Censure with
'Indignation and Triumph. And yet after all,'
'I find it had been but too true a Charge.——
'O! how hard a Thing is it to know myself?
'——This, like all other Knowledge, the more
'I have of it, the more sensible I am of my Want
'of it *.'

The Difficulty of Self-government and Self-possession arises from the Difficulty of a thorough Self-acquaintance which is necessary to it. I say a *thorough* Self-acquaintance, such as has been already set forth in its several Branches, (Part I.) For as Self-government is simply impossible (I mean considered as a Virtue) where Self-ignorance

---

* *Cicero* was without Doubt the vainest Man in Life; or he never could have the Face to beseech *Cocceius*, in writing the Roman History, to set the Administrations of his Consulship in the most distinguished Point of Glory, even at the Expence of Historical Truth; and yet when he is begging a Favour of the like Kind even of *Cato* himself, he has these astonishing Words.—— Si quisquam fuit unquam remotus et naturâ et magis etiam (ut mihi quidem sentire videor) ratione atque doctrinâ ab INANI LAUDE ET SERMONIBUS VULGI, ego profectò is sum. lib. 15. Ep. 4. *If ever any Man was a* STRANGER TO VAIN GLORY, *and the Desire of Popular Applause, it is myself*; *and this Disposition which I have by Nature, is (methinks) grown yet stronger by Reason and Philosophy.*——Ah! how secretly doth Self-ignorance (not only insinuate into, but) conceal itself within the most improved and best cultivated Minds!——Reader, beware.

prevails, so the Difficulty of it will decrease in Proportion to the Degree in which Self-acquaintance improves.

Many, perhaps, may be ready to think this a Paradox; and imagine that they know their predominant Passions and Foibles very well, but still find it extremely difficult to correct them. But let them examine this Point again, and perhaps they may find, that *that* Difficulty arises either from their *Defect* of Self-Knowledge, (for it is in this as in other Kinds of Knowledge, wherein some are very ready to think themselves much greater Proficients than they are) or else from their *Neglect* to put in Practice that Degree of Self-Knowledge they have. They know their particular Failings, yet will not guard against the immediate Temptations to them. And they are often betrayed into the immediate Temptations which overcome them, because they are ignorant of, or do not guard against, the more *remote* Temptations, which lead them into those which are more immediate and dangerous, which may not improperly be called the Temptations to Temptations; in observing and guarding against which, consists a very necessary Part of Self-Knowledge, and the great Art of keeping clear of Danger, which, in our present State of Frailty, is the best Means of keeping clear of Sin.

To correct what is amiss, and to improve what is good in us, is supposed to be our hearty Desire, and the great End of all our Self-research.

research. But if we do not endeavour after this, all our Labour after Self-Knowledge will be in vain. Nay, if we do not endeavour it, we cannot be said heartily to desire it. 'For there is most of the Heart, where there is most of the Will; and there is most of the Will, where there is most Endeavour; and where there is most Endeavour, there is *generally* most Success. So that Endeavour must prove the Truth of our Desire, and Success will generally prove the Sincerity of our Endeavour (*f*).' This, I think, we may safely say, without attributing too much to the Power of the Human Will, considering that we are rational and free Agents, and considering what effectual Assistance is offered to them who seek it, to render their Endeavours successful if they are sincere. Which introduces the Subject of the following Chapter.

CHAP. X.

Fervent and frequent Prayer the most effectual Means for attaining true Self-Knowledge.

LASTLY, *the last Means to Self-Knowledge which I shall mention is, frequent and devout Applications to the Fountain of Light, and the Father of our Spirits, to assist us in this important Study, and give us the* true Knowledge of ourselves.
                                                                    This

(*f*) *Baxter.*

This I mention laſt, not as the leaſt, but, on the contrary, as the greateſt and beſt Means of all, to attain a right and thorough Knowledge of ourſelves: and the Way to render all the Reſt effectual. And therefore, though it be the laſt Means mentioned, it is the firſt that ſhould be uſed.

Would we know ourſelves, we muſt often converſe not only with ourſelves in Meditation, but with GOD in Prayer. In the loweſt Proſtration of Soul, beſeeching the Father of our Spirits to diſcover them to us; *in whoſe Light we may ſee Light,* where before there was nothing but Darkneſs; to make known to us the Depth and Devices of our Heart. For without the Grace and Influence of his Divine Illuminations and Inſtructions, our Hearts will, after all our Care and Pains to know them, moſt certainly deceive us. And Self-love will ſo prejudice the Underſtanding, as to keep us ſtill in Self-ignorance.

The firſt Thing we are to do in order to Self-Knowledge is, to aſſure ourſelves that our Hearts *are deceitful above all Things.* And the next is, to remember that *the Lord ſearcheth the Hearts, and trieth the Reins* *, *i. e.* that He, the (Καρδιογνωσης) *Scarcher of all Hearts* †, hath a perfect Knowledge of them, deceitful as they are. Which Conſideration, as it ſuggeſteth to us the ſtrongeſt Motive to induce us to labour after a true

---

* *Jer.* xvii. 10.   † 1 *Chron.* xxviii. 9.

a true Knowledge of them ourselves; so it directs us at the same Time how we may attain this Knowledge; *viz.* by a humble and importunate Application to Him, to whom *alone* they are known, to make them known to us. And this, by the free and near Access which his Holy Spirit hath to our Spirits, he can effectually do various Ways; *viz.* by fixing our Attentions; by quickning our Apprehensions; removing our Prejudices, (which, like a false Medium before the Eye of the Mind, prevents its seeing Things in a just and proper Light;) by mortifying our Pride; strengthening the intellective and reflecting Faculties; and enforcing upon the Mind a lively Sense and Knowledge of its greatest Happiness and Duty; and so awakening the Soul from that carnal Security and Indifference about its best Interests, into which a too serious Attention to the World is apt to betray it.

Besides, Prayer is a very proper Expedient for attaining *Self-Knowledge*, as the actual Engagement of the Mind in this devotional Exercise is in itself a great Help to it. For the Mind is never in a better Frame, than when it is intently and devoutly engaged in this Duty, It has then the best Apprehensions of GOD, the truest Notions of itself, and the justest Sentiments of earthly Things; the clearest Conceptions of its own Weakness, and the deepest Sense of its own Vileness; and consequently is

in

in the best Disposition that can be, to receive a true and right Knowledge of itself.

And, Oh! could we but always think of ourselves in such a Manner, or could we but always be in a Disposition to think of ourselves in such a Manner, as we sometimes do in the Fervour of our Humiliations before the Throne of Grace, how great a Progress should we soon make in this important Science? Which evidently shews the Necessity of such devout and humble Engagements of the Soul, and how happy a Means they are to attain a just *Self-acquaintance.*

AND NOW, Reader, whoever thou art that has taken the Pains to peruse these Sheets, whatever be thy Circumstances or Condition in the World, whatever thy Capacity or Understanding, whatever thy Occupations and Engagements, whatever thy favourite Sentiments and Principles, or whatever Religious Sect or Party thou espousest, know for certain, that thou hast been deeply interested in what thou hast been reading; whether thou hast attended to it or no. For it is of no less Concern to thee than the Security of thy Peace, and Usefulness in this World, and thy Happiness in another; and relates to all thy Interests, both as a Man and a Christian. ——— Perhaps thou hast seen something of thine own Image in the
Glass

Glass that has now been held up to thee. And wilt thou go away, and soon *forget what Manner of Person thou art?* —— Perhaps, thou hast met with some Things thou dost not well understand or approve. But shall that take off thine Attention from those Things thou dost understand and approve, and art convinced of the Necessity of?——If thou hast received no Improvement, no Benefit from this plain practical Treatise thou hast now perused; read it over again. The same Thought, you know, often impresses one more at one Time than another. And we sometimes receive more Knowledge and Profit by the second Perusal of a Book than by the first. And I would fain hope that thou wilt find something in this that may set thy Thoughts on Work, and which, by the Blessing of GOD, may make thee more observant of thy Heart and Conduct; and in Consequence of that a more solid, serious, wise, established Christian.

But will you, after all, deal by this Book you have now read, as you have dealt by many Sermons you have heard? Pass your Judgment upon it according to your received and established Set of Notions; and condemn or applaud it, only as it is agreeable or disagreeable to *them*; and commend or censure it, only as it suits or does not suit your particular *Taste*; without attending to the real Weight, Importance, and Necessity of the Subject abstracted from those Views; Or will you be barely content with the

Enter-

Entertainment and Satisfaction, which some Parts of it may possibly have given you; to assent to the Importance of the Subject, the Justness of the Sentiment, or the Propriety of some of the Observations you have been reading; and so dismiss all without any further Concern about the Matter? ——— Believe it, O Christian Reader, if this be all the Advantage you gain by it, it were scarce worth while to have confined yourself so long to the Perusal of it. It has aimed, it has sincerely aimed, to do you a much greater Benefit; to bring you to a better Acquaintance with one you express a particular Regard for, and who is capable of being the best Friend, or the worst Enemy, you have in the World; and that is *yourself*.——— It was designed to convince you, that would you live and act consistently, either as a Man, or a Christian, you must *know yourself*; and to persuade you under the Influence of the foregoing Motives, and by the Help of the fore-mentioned Directions, to make *Self-Knowledge* the great Study, and *Self-government* the great Business of your Life. In which Resolution may Almighty GOD confirm you; and in which great Business may his Grace assist you, against all future Discouragements and Distractions? With Him I leave the Success of the whole; to whom be Glory and Praise for ever.

F I N I S.

www.ingramcontent.com/pod-product-compliance
Lightning Source LLC
Chambersburg PA
CBHW022013220426
43663CB00007B/1061